THE SYNCHRONICITY GUIDEBOOK

HIGH-TECH MEDITATION

AMETHYST PUBLISHING

Synchronicity Foundation, Inc.
P. O. Box 694
Nellysford, VA 22958

Toll Free: 1-800-962-2033
In Va: (804) 361-2323
9am - 6pm Eastern Time Daily

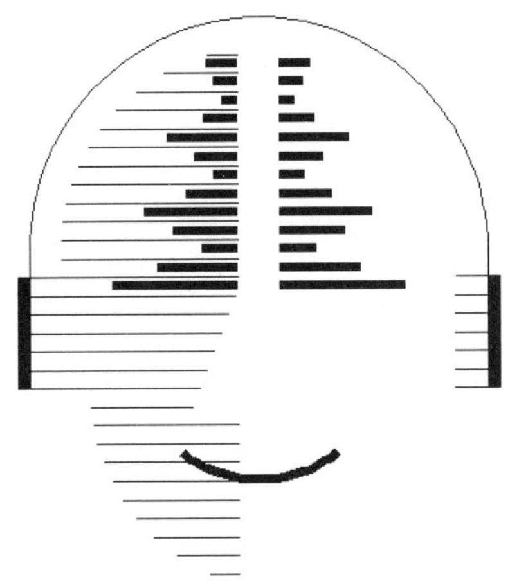

THE SYNCHRONICITY GUIDEBOOK...
HIGH-TECH MEDITATION

Based on the work of Master Charles and the Synchronicity Foundation

Compiled By: Cynthia Larsen and Paul Shannon

Amethyst Publishing

AMETHYST PUBLISHING
Route 1, Box 192-B
Faber, Virginia 22938

Library of Congress Catalog Number 93-074177
ISBN Number 1-884068-20-0

**This Book Is Dedicated
To Mastership**

TABLE OF CONTENTS

8 Preface

16 Part One: Contemporary Meditation for Westerners

The mechanics of meditation, classical and contemporary.

32 Part Two: Meditation for a New Age

How High-Tech Meditation works and how it differs from other products in the "mind toys" marketplace.

58 Part Three: How To Practice High-Tech Meditation

Guidelines for establishing a regular meditative practice.

74 Part Four: What To Expect From High-Tech Meditation

How and why Synchronicity High-Tech Meditation transforms the lives of its users.

98 Part Five: Mastership: a Contemporary Understanding

Explanation of the transformational power of mastership in contemporary terms. The role of the master in the human evolutionary journey.

108 Epilogue

PREFACE

There are many paths of meditation. Bookstores and libraries are filled with tomes describing the ancient spiritual practices of various meditative traditions. Buddhist, Taoist, Sufi, Tantric, Hindu- and their various accoutrements- mandalas, beads, chants, mantras. It would seem one needs several degrees in Eastern philosophy to begin to understand them all. There is, however, one book which describes the contemporary Western application of such ancient meditative techniques and this is it.

The Synchronicity Experience, popularly known as High-Tech Meditation, was originated by

Master Charles, a contemporary American non-sectarian spiritual educator. As a close disciple of the renowned spiritual master, Paramahansa Muktananda, Master Charles spent many years in India and was initiated as a Vedic monk.

In 1983, following Muktananda's death, he relinquished the orthodox lifestyle and returned to America in order to bridge the gap between contemporary Western society and the spiritual culture of the East. Based on a lifetime of spiritual exploration and extensive experience in music and sound production, as well as a thorough knowledge of comparative religion and philosophy, Master Charles has created a system which combines classical Eastern philosophic principles with Western contemporary empirical technology.

This work was first publicly formulated and presented in 1983 in the Synchronicity Paradigm. It was here that he revealed that very specific frequencies of whole brain function could be entrained in the human brain through intricate sound patterning, resulting in conscious access to the subconscious and unconscious databanks of the brain-mind computer. Precision rescripting could then be accomplished through the use of subliminal affirmations, resulting in the progressive development of constancy in expanded states of consciousness awareness. In this context, the ancient process of enlightenment is viewed as behavioral modification.

Synchronicity High-Tech Meditation has been a great success for one simple reason - it works. It is precise and efficient, taking far less time and effort than traditional meditation systems. This acceleration factor is one of the principal differences between Master Charles' Synchronicity Holodynamic Technology (High-Tech Meditation) and traditional orthodox (low-tech) systems of meditation. The speed with which measurable changes occur in one's states of awareness has been confirmed by measurement of the brain-wave patterns produced by regular users of Synchronicity Meditative Technology. These were compared to the measurements produced by those using traditional meditative systems. Our findings empirically validate the assertion first advanced by Master Charles in the Synchronicity Paradigm, that Synchronicity Technology yields a fourfold acceleration factor over classical methods of meditation.

Master Charles' seeming abandonment of traditional, orthodox religious contexts in favor of the contemporary high-tech approach has made him one of the most innovative spiritual educators of our times.

In his view, one need no longer spend fifty years meditating in a cave to attain enlightenment. Through the use of modern technology, one can have the cave experience without leaving the comfort of home - and in much less time. As Master Charles puts it, “Why use a horse-drawn carriage when you can have a modern automobile?"

THE SYNCHRONICITY GUIDEBOOK has been designed for the general reader and does not require a background in either Eastern philosophy or Western science. Yet those who have experience in these areas will find that the work of Master Charles and the Synchronicity Foundation brings unprecedented clarity to the often mystifying subject of meditation. Where appropriate, correspondences between classical and contemporary High-Tech Meditation have been made, in order to assist the reader in understanding how the two systems complement and validate each other.

The text of this book is divided into five main sections:

Part One: “Contemporary Meditation for Westerners” attempts to explain, in words, the mechanics of states of awareness which exist beyond words, form or thought.

Part Two: "Meditation For A New Age" describes the technology behind the technology. It explains what High-Tech Meditation is, how it works and how it differs from other brain tapes and machines in the 'mind toys' marketplace.

Part Three: "How To Practice High-Tech Meditation" provides simple guidelines for establishing a regular meditative practice and offers recommendations on various support systems which will help you get the most out of your meditation.

Part Four: "What To Expect From High-Tech Meditation" offers revealing insights into how Synchronicity High-Tech Meditation technology invariably transforms the lives of regular users. Not only will you learn what happens as a result of High-Tech Meditation, but you'll understand why.

Part Five: "Mastership: A Contemporary Understanding" explains the transformative power of mastership in contemporary Western terms and discusses the all-important role of the master in the human evolutionary journey.

Throughout the book, set off from the rest of the text, are quotes from Master Charles. With dazzling precision and wit, he brings a clarity to the mind-bending subject of meditation that only true enlightening mastership can provide.

Master Charles is brilliant, charismatic, outrageous, amusing, astounding and endearing. His words may affect you deeply. Whatever your reaction, remember that what you are responding to is but a reflection of yourself.

***"Someone as outrageous as myself
lives simply to demonstrate to you that,
yes, indeed,
there IS a truthful reality,
a Source, a God,
and what's more, it is also YOU,
which is both the greatest mystery
and the greatest joke
in all the universes."***

The material contained in this book provides a fascinating glimpse into the work being done by the "spiritual pioneers" who make up the Synchronicity Foundation. We are a curious, eclectic group of individuals, brought together by our willingness to challenge established boundaries of personality and of awareness. And whether you are a seasoned meditator or a beginner, you are most welcome in our world.

The urge to know one's Source, to understand where we came from and where we are going has always been part of the human journey. We now have the technology at our disposal which can help answer these questions with a precision and efficiency never before available. Synchronicity High-Tech Meditation is very simply the contemporization of the journey of enlightenment. It is for contemporary seekers of Truth that this book has been written.

PART ONE:

CONTEMPORARY MEDITATION FOR WESTERNERS

WHAT IS MEDITATION?

Webster's Dictionary defines meditation as "deep, continued thought." Should you really wish to learn to meditate, we suggest you avoid Mr. Webster - for, in fact, the goal of all meditation systems, classical or contemporary, is the relinquishment of thought.

Classical meditative systems hold that true meditation occurs whenever the mind transcends the illusion of duality, i.e., the false perception that there is an "other". At this point, the stilling, or equilibration, of the mind opens the meditator to the non-dual experience of true Reality as one universal consciousness.

This state, which Master Charles refers to as Source Consciousness Awareness, has been the goal of all meditative systems from time immemorial.

The true meditative experience is one of pure, Sourceful awareness, in which dualistic thought is non-existent. It is the complete identification of oneself with a euphoric, ecstatic, all-inclusive, eternal consciousness. While such an expanded state of awareness may be more than most Westerners expect when they sit for meditation, the fact is that, by modifying the behavior of the mind, one is able to achieve a "super-conscious" state of expanded awareness in which the meditator recognizes that ALL IS ONE.

"In the true meditative state of awareness, there is the recognition that all existence, all life, is but the play of ONE consciousness, ONE reality, ONE Source celebrating itself as all and everything."

Traditionally, two major meditative techniques have been used to still the thought processes and entrain attention. Through the first of these techniques, known as concentration, the meditator focuses attention on a single object. Typical focusing devices include the breath, a mantra (non-dual affirmation or prayer), a candle flame, or a circular object of art known as a mandala. When complete concentration of focus is achieved, all mental distraction ceases and the meditator's awareness merges with the object of meditation.

With the second technique, that of mindfulness, focused attention is also essential. In this case, a full ***awareness*** of each successive moment of the meditator's experience is required. The meditator becomes the detached observer of his own thoughts, and eventually becomes aware of the most minute fluctuations in his awareness. When mental processes are observed with continuous detachment, they gradually disappear, thus opening the way to expanded states of awareness.

The nature of the expanded states achieved through the practice of either concentration or mindfulness differ. The path of concentration can lead to states of ultra-subtle perception in which awareness all but disappears. This state marks the outer limits of human perception. For advanced practitioners of the path of mindfulness, however, a state of awareness arises in which all physical and mental phenomena cease entirely. This is the experience which the Buddhists call no-mind. The initial entry into this state may be considered to be the meditator's "awakening". Subsequent insights resulting from the repeated experience of no-mind lead to "liberation".

It is at this point in the enlightening process that the meditator transcends human limitations and becomes an enlightening human being, fully identified with non-dual Source. Historically, it was the promise of such enlightening awareness that led spiritual seekers to twist themselves up like pretzels and spend fifty years meditating in a cave. Contemporary seekers have discovered a faster, easier and far more precise way of achieving the same end.

HIGH-TECH MEDITATION

The emergence of High-Tech Meditation is a predictable and appropriate outgrowth of the human evolutionary journey. It is natural for us, as contemporary Westerners, to gravitate toward any technology which allows us to expand our understanding of ourselves and our environment.

The development and use of Western technology applied to the ancient practice of meditation is an indication that Western culture is moving towards a fuller, more balanced understanding of the transcendental nature of reality. To the best of our ability, we are using our scientific understanding to help unfold the secrets of the universe. In the case of meditation, the traditional techniques seem far less precise and efficient when compared to the accelerated growth resulting from precision High-Tech Meditation. This union of the objective left-brain world with the subjective right-brain world is a means of creating the balance essential for true meditation. Up until now, such union has been noticeably lacking in our left-brain dominant Western culture.

As with traditional meditative techniques, High-Tech Meditation requires that one devote a regular period of time each day to formal sitting meditation. The reason for this is that the mind must be freed from the necessity of monitoring the never-ending parade of stimuli that results from movement either within the body or within the environment. When we examine what happens in the process of sitting meditation, we find that meditation is an entrainment into a state of being, a state of awareness and clear perception that comes when phenomenal experience is superceded by a 'still' state of being.

What occurs at this point is most unexpected, for a totally new world opens to the meditator. One becomes aware of a dynamic and powerful multi-dimensional energy which profoundly influences all aspects of the meditator's awareness. There is much to explore and experience within this new state. Once one has entered the realm of the subtler dimensions, life will never again be the same.

Initially, this new, expanded state of awareness is a fragile thing which disappears quickly with any disturbance. However, with continued meditative practice, this clear, pure, intoxicating, transcendent state becomes a more constant experience.

It is at this point that contemporary meditators will usually "happen upon", synchronistically speaking, the technique of High-Tech Meditation. It is common for experienced meditators to find that they have reached a plateau in their state of awareness beyond which they cannot go on their own. It is then that one requires a push, an acceleration, into the deeper dimensions of one's own perception and awareness. High-Tech Meditation is an advanced meditative discipline capable of catalyzing awareness on to ever greater levels of expansion.

For those with no previous meditative experience, High-Tech Meditation provides an impactful initiation into the art and science of awareness expansion. Without a background in the traditional orthodoxy of meditative experience, the beginning High-Tech meditator will discover the world of meditation in a whole new way, never before known in human experience. Is it the same world as its ancient counterpart? The answer is yes and no.

THE EMPIRICAL CONTEXT OF MEDITATION

The classical Eastern understanding of reality as One Consciousness which exists beyond the dualistic limits of the mind is supported by contemporary science. In an empirical context, quantum physics and the new science assert that all phenomena within our universe are actually parts of one all-encompassing organic pattern or hologram. Nothing exists independently, in spite of appearances to the contrary. The philosophical implications of quantum mechanics support the classical view that everything is but the play of one reality consciousness.

The Holographic Theory, the Super Strings Theory, the Identity Theory, Einstein's vision of the universe and the Big Bang Theory all support the reality of one energy polarized into a seeming duality. When duality disappears through equilibration, only one reality remains. Each of these theories contributes to the recognition that, on the most fundamental level, our universe is one indivisible Source. On the microcosmic level, as demonstrated within the structure of the human brain, the illusion of duality is exemplified by the polarization, or lateralization, of the right and left hemispheres of the brain. Synchronization, or the coherent equilibration of the brain hemispheres, leads to the expansion of awareness which results in the relinquishment of the illusion of duality.

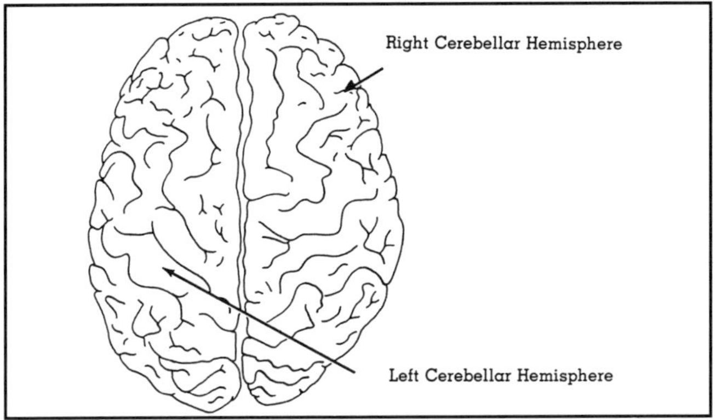

Figure 1. Top view of the brain illustrating left and right hemispheres.

According to contemporary science, what determines synchronization, or the degree of coherent balance of brain function, is thought. It is an empirical fact that thought can manipulate the subtlest of electromagnetic energy fields. As such, thought constructs determine the interactions within the brain that result in either whole or fragmented brain function. Hence, dualistic thought is a disturbance or tension which accelerates energy frequency of vibration. This, in turn, upsets balance and prevents the experience of the truthful awareness that all is Source.

In the traditional Eastern philosophical understanding, this limiting, dualistic data consists of stored impressions contained within the subtle bodies of the human form. These bodies transmigrate with the soul through many lifetimes. In contemporary terms, such "bodies" may be regarded as energy fields. The stored impressions constitute what we understand to be the subconscious/unconscious data banks which comprise the memory storage areas of the brain-mind computer. This subconscious/unconscious data, which is encoded in the individual energy field, contains all of the information recorded throughout the individual's multidimensional experience. The print-out of such data, both individual and collective, becomes our conscious world. Simply put, our reality is but a projection of what we believe reality to be.

In our heavily left-brain dominant Western culture, most people's thoughts are governed by the rational, linear, dualistic, negative thinking characteristic of the left brain. Such extreme hemispheric imbalance severely restricts whole brain function as demonstrated by recent studies which have indicated that average humans use no more than 5%-10% of their total brain capacity. This limited brain function is a direct result of the extreme fragmentation characteristic of the beta state. This state of extremely lateralized imbalance and limited brain function is dominated by the active beta brain-wave frequencies commonly found in the waking state. Such beta dominance affects far more than just our brain-wave pattern. In practical terms, beta dominance results in stress, which often leads to adverse effects in a variety of areas including work, health and relationships.

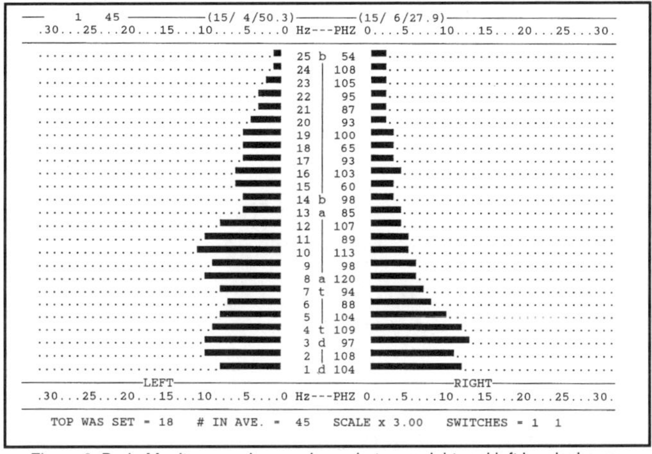

Figure 2. Brain Monitor reveals asynchrony between right and left hemispheres of the brain.

As one practices meditation, brain-wave frequencies decelerate into the more relaxed alpha frequencies and hemispheric balance improves. As whole brain function increases, there is a corresponding reduction in stress, and an increase in focus, concentration and feelings of health and well-being. Whole brain function improves even more as one decelerates further into the deeply meditative and creative theta brain-wave frequencies. It is in the relaxed lower alpha and theta frequencies that access to subconscious data is achieved.

"In the state of synchronicity, the stress of polar opposites is relinquished and with it all dualistic illusion. This is the state of pure balance in which human beings experience themselves as Source."

The theta/delta brain-wave frequencies are the deepest and slowest of brain frequencies and it is at this level that the unconscious is accessed. At the deepest end of the delta range is the threshold to non-dual awareness or witness consciousness. It is at this point of transcendental access that paranormal phenomena or parapsychological manifestations occur spontaneously as a natural adjunct to the state of whole brain synchrony.

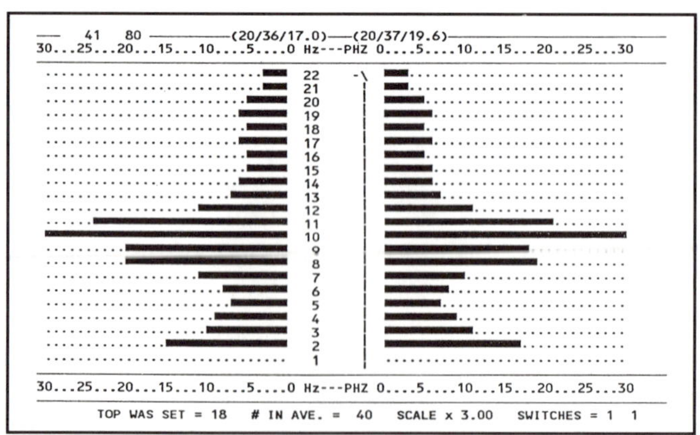

Figure 3. An excellent example of synchrony and symmetry between the right and left hemispheres of the brain. Perfect synchronization is indicated by the absence of (+) or (-) marks in the center column.

As synchronization increases, the brain's natural opiates, called endorphins, are released. The endorphins act as natural narcotics and are the body's way of reacting to pain or stress. When secreted in abundance, as occurs during deep meditation, there arises a deeply pleasurable state of awareness which is sometimes referred to as "God intoxication" or "transcendental bliss". Human beings instinctively seek the intoxicated state, which is characterized by feelings of peace, spontaneity and blissful euphoria.

Empirically speaking, this is validated by the sleep cycle. During sleep the brain goes into delta frequencies, releasing maximum opiates and heavily endorphinating the brain. That sleep is pleasurable is undeniable. So is the fact that human beings enjoy becoming intoxicated. But aside from the obvious damage done by the residues they leave behind, the problem with synthetically-induced highs achieved through the use of drugs or alcohol is that the elevated state they produce is neither conscious nor continuous. In the advanced meditator, however, blissful intoxication is conscious and constant, both of which are necessary for enlightenment to occur. Throughout history the human journey has been intimately linked to the pursuit of such states of intoxification or elevation, the "high" of life lived Sourcefully, without stress.

"I say to you, 'stay high',
give up everything else,
relinquish it all and just stay high.
High in the identity of the One
that you are.
All else is nonsense
and it will only make for you
a very weighty coffin."

Earlier in this section we posed the question: Is the meditative experience achieved through High-Tech Meditation the same as its counterpart? Based upon what has been presented so far, we have discovered that yes, the meditative state accessed through Synchronicity Technology is the same as that achieved with classical meditative systems. However, as we explore exactly how High-Tech Meditation works, we will find that for the contemporary High-Tech meditator the expansion of awareness unfolds much more rapidly due to the precision technology. This is what Master Charles refers to when he speaks of the difference between driving a horse-drawn carriage and a modern automobile.

In the following section we will become familiar with the beta, alpha, theta and delta brain-wave frequencies. We will examine how Synchronicity High-Tech Meditation uses sound vibration to affect the functioning of the human brain and how hemispheric synchronization is accomplished and maintained. We will consider the effect of subliminal affirmation upon the data contained within the subconscious/unconscious data banks, which we refer to as the human brain-mind computer. Based on an intelligent understanding of the effects of this technology on the evolution of human consciousness, it is apparent that we have arrived at a new age of enlightening evolution.

PART TWO:

MEDITATION FOR A NEW AGE

BRAIN WAVES AND MEDITATION

What specifically do the terms "beta", "alpha", "theta", and “delta” mean for the High-Tech meditator? What type of changes actually occur in brain-wave frequencies as one progressively enters deeper states of meditation? Current technology, including the one-of-a-kind "Brain Monitor" developed by Synchronicity's research team, enables us to answer these questions with precision. In this section we will consider how Synchronicity's empirical sound technology affects the development of hemispheric synchronization and whole brain function, resulting in a transformational experience which unfolds from alpha through delta, from light relaxation to deep transcendental meditation.

ALPHA

When brain activity is focused primarily within the 8-13 Hertz (cycles per second) range, one experiences what is termed an "alpha" state of awareness. Alpha brain waves are produced by most people when they concentrate or focus the mind, or relax by sitting or reclining with their eyes closed. Though alpha brain waves may be found throughout the brain during these times, they are mainly localized over the motor and visual cortices. Alpha brain waves are very often produced in bursts (trains of waves) or pulses (single waves), but some people, and especially High-Tech meditators, tend to produce continuous trains of alpha waves. When your brain is producing alpha frequencies, your experience is one of mild relaxation; you experience a pleasant, reasonably alert awareness.

This experience is often taken for granted by meditators, yet the state of alpha represents a relatively stress-free and euphoric state of being. The general understanding is that the more alpha that is produced in ordinary states of awareness, the easier it is to access deeper meditative states.

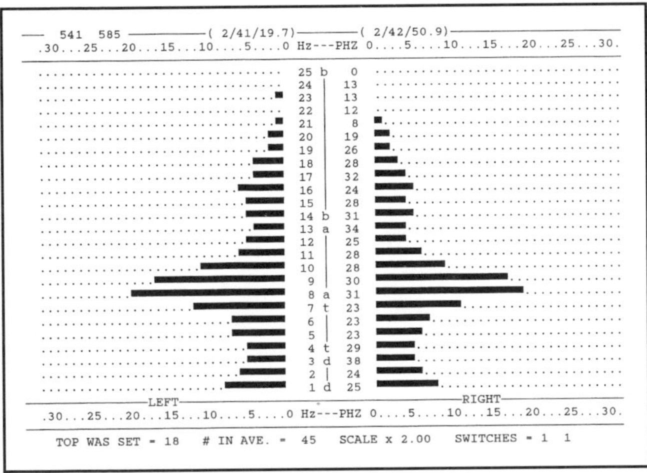

Figure 4. A good example of alpha production with dominant peak at 8 Hz.

Practically speaking, individuals such as artists or photographers, who use the visual and spatial abilities characteristic of the right brain, seem to produce alpha more easily than cognitive, left-brain thinkers. This is due to the balancing of the right and left brain hemispheres, resulting in increased synchrony. In the same way, practiced High-Tech meditators, who have been repetitively patterned and become familiar with alpha brainwave production through the use of this Synchronicity meditative technology, can produce alpha frequencies readily.

By looking at an individual's alpha production, it is possible to determine not only if the person is a meditator, but also the length of time the individual has been practicing meditation. Beginners produce alpha in the 10-12 Hz. range, while experienced meditators consistently produce low frequency alpha in the region of 7-9 Hz.

THETA

"Theta" brain waves are produced in the frequency range from 3.5-7 Hz., which are slower brain waves than alpha waves. Like alpha, theta is characterized by a blissful sense of well-being. However, in theta the experience of expansion increases. Theta is considered to represent the synchronized state of awareness in which creativity and imagery predominate. Experienced meditators are accustomed to a wide variety of "inner" images and visions. In High-Tech Meditation, these images do indeed seem correlated with increased amounts of theta brain waves induced by the sound patterning technology.

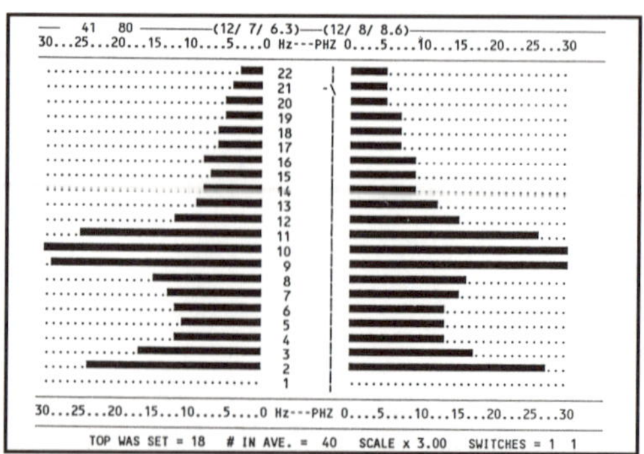

Figure 5. While theta waves never achieve the amplitude of alpha waves, this individual demonstrates excellent theta production between 3.5 and 7 Hz.

Perhaps of greatest importance with regard to alpha and theta frequencies is the association of lower alpha/upper theta brain-wave production with what appears to be a "doorway", or entry, into the subconscious. The technology of High-Tech Meditation has enabled us to explore this dimension of conscious experience thanks to the innovative genius of its creator, Master Charles, who was the first to describe the specific relationship between brain-wave frequencies and the subconscious and unconscious data banks of the brain/mind computer.

Becoming familiar with the experience of theta frequency brain-wave production through High-Tech Meditation enables one to remain fully conscious during the time when theta frequencies are present. With experience, it is possible to recreate or remain within this blissful experience for an extended period. During these times, the conceptual elements of subconscious and unconscious experience or behavior are brought to a conscious level of awareness, providing ready access to these realms. This conscious access to incongruent or antiquated data is one of the key factors in releasing the conceptual, thought-form limitations which prevent one from experiencing whole brain synchrony and the state of Source Consciousness Awareness.

This may sound radical and in a way it is. Yet in reality, High-Tech Meditation is simply an acceleration of the mind's natural method of data processing. Consciousness is structured in such a way that the subconscious and unconscious normally only reveal what one can integrate at any given time. When the current level of subconscious/unconscious data has been integrated, the process of unfolding continues, revealing still deeper levels of subconscious and unconscious data. This method of data processing is essentially the same as what goes on in the course of everyday life. High-Tech Meditation simply brings a much greater degree of precision, and therefore efficiency, to what is a normal process of growth and evolution. As a direct result, there is also a much greater degree of awareness.

DELTA

The states where "delta" frequencies are produced are the most challenging for the High-Tech meditator. Delta brain waves are generally associated with deep sleep and are the slowest of the brain waves, being from 0-3.5 Hertz. Usually only the most advanced meditators can remain conscious while producing delta brain waves. The experience leading to the threshold of "no mind" occurs in the deepest levels of the delta brain-wave frequencies. In this deeply decelerated state, the window to the unconscious can be held open. This allows the meditator to relinquish the most rudimentary dualistic data and to remain continuously in very elevated, transcendental, non-dual states of awareness.

Our research has shown that with the ongoing entrainment provided by regular use of Synchronicity delta level tapes, High-Tech meditators develop constancy in the production of delta brain waves. Thus, after only a few years of usage, High-Tech meditators produce brain wave patterns equal to those individuals who have spent a lifetime of practice using traditional methods.

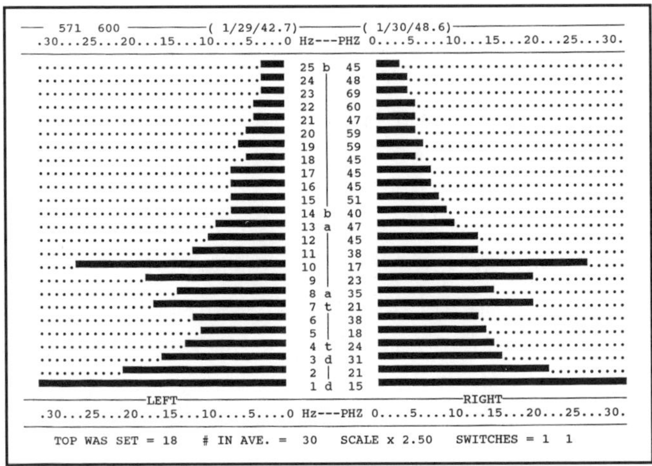

Figure 6. An example of high amplitude delta production during the waking state.

BETA

"Beta" brain waves are the most rapid of the brain-wave frequencies. They are characterized by frequencies ranging from 14-40 Hertz. Their low amplitude, high frequency wave form is characteristic of a wakeful, aroused and thinking brain. The production of beta brain waves, and nothing else, usually means a relatively stressful, active mental state, one not especially conducive to transcendental levels of meditative experience. Yet practiced meditators should retain proportional beta function but show higher levels of alpha, theta and even some delta waves occurring at the same time. If no beta is present, the subject would be considered to be in Stage II or III sleep, depending on the ratios of the other frequencies.

The complete, whole brain experience so desired by meditators does require some beta to retain awareness. However, in general, most people experience beta as a tiring, stress-inducing state and relinquish its dominance during times of relaxation and meditation. Meditation's well-deserved reputation as a stress-reliever comes from the fact that it induces alpha frequencies which offset the stressful effects of the beta frequencies.

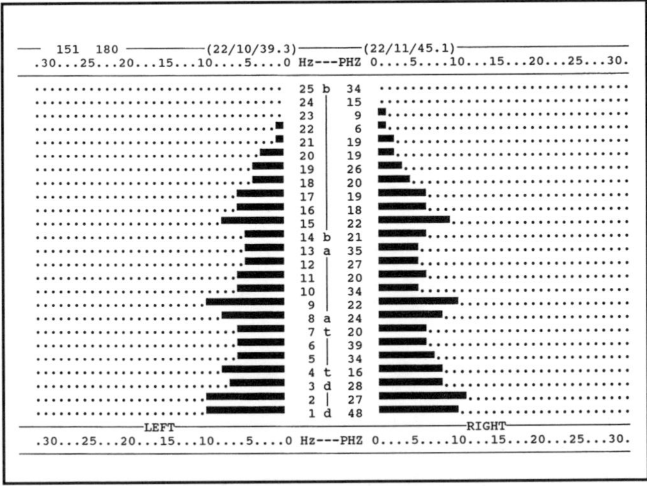

Figure 7. Brain Monitor shows dominant beta peak at 15 Hz.

Assessment of brain function in terms of beta, alpha, theta and delta brain wave production is an important part of a meditator's understanding in today's contemporary transformational environment. Synchronicity High-Tech Meditation has taken straightforward EEG procedures out of their restrictive clinical domain and made the information from them more accessible. It is advisable for High-Tech meditators to become familiar with their own brain wave "fingerprint", for the information available with these scans serves to de-mystify the normal functioning of this most important physiological counterpart to the meditative experience, the human brain.

THE SYNCHRONICITY PARADIGM

Synchronicity High-Tech Meditation, with its precise and efficient ability to create increases in hemispheric synchrony, goes much further than simply entraining the brain into alpha, theta and delta brain wave functions. As meditation deepens from ordinary waking consciousness into transcendental states of awareness, the brain waves decelerate in frequency from predominantly beta (14-40 Hertz) to include alpha (8-13 Hertz) and theta (3.5-7 Hertz) and eventually delta (0-3.5 Hertz) frequencies. The subconscious begins to release data in the lower alpha and upper theta frequency ranges, allowing one to become consciously aware of one's own incongruent data. As the brain waves deepen into the lower theta and delta frequencies, the unconscious begins to systematically unfold the information present within the more primal dimensions of the collective awareness.

"Transformation in the individual is simultaneous in the universal."

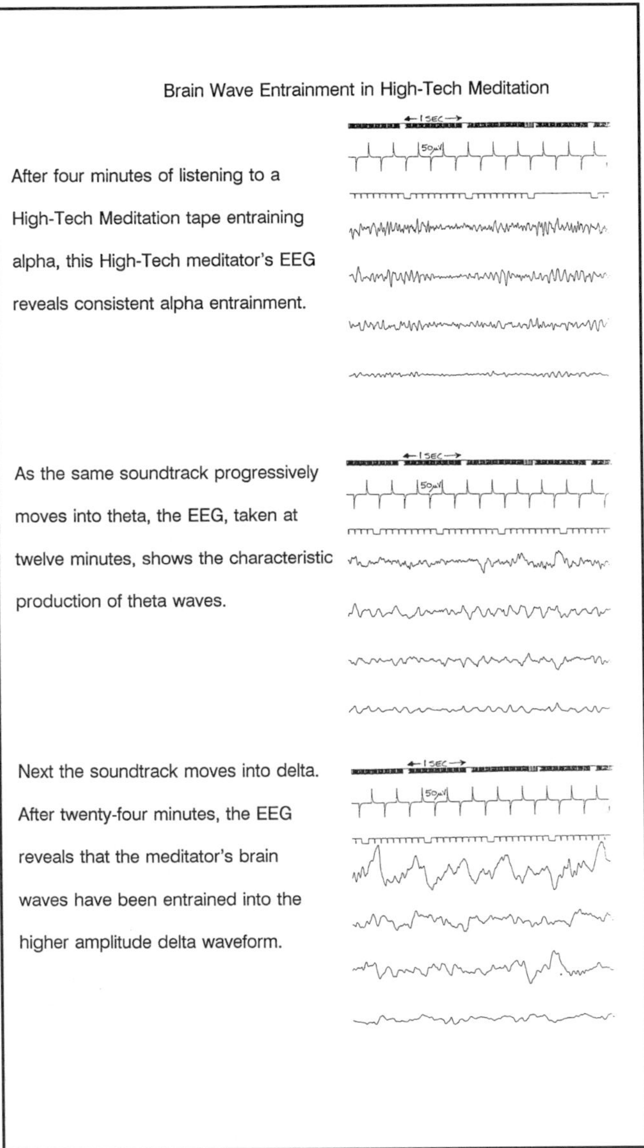

Figure 8. EEG reading reveals how Synchronicity Technology progressively entrains brain waves to slower and deeper frequencies.

Master Charles originated and first introduced this understanding of the multidimensional brain-mind function in the Synchronicity Paradigm, which forms the basis of the Synchronicity Experience in all its applications. When explaining the process by which the brain evolves to greater levels of clarity, Master Charles characterizes the work of Nobel Prize-winning scientist, Ilya Prigogine, as a five-fold cycle which includes peak, breakthrough, illumination, catharsis and integration. These steps directly correspond to the clearing and expansion which occur as a result of meditation and may be understood as follows:

When the brain reaches its maximum, or peak, level of synchrony, it suspends activity. A breakthrough into new levels of Sourceful clarity is experienced. A resultant upheaval in the subconscious and unconscious data banks takes place as old, incongruent and conflicting data are released prior to full integration of the new experience. This cycle is repeated over and over, leading to increasing levels of clarity and expansion in awareness. Thus, the human evolutionary journey may be seen as an unfolding process which always reveals a progression from the dual to the non-dual.

Through the practice of High-Tech Meditation, the data released from both the subconscious and unconscious may be observed and cleared.

This is what is referred to as integration. An analogy is that of a light being cast on areas which are kept "dark" to everyday awareness. With the light, the darkness is dispelled and fuller dimensions of one's experience are perceived. With High-Tech Meditation, this emerging data is either cleared, that is, removed completely, or it is brought into conscious awareness and integrated without repeated repression.

High-Tech Meditation brings precision and efficiency to the five-fold evolutionary cycle and has given us, for the first time, an empirical understanding of the dynamics of the meditative process. Through repeated access to the deepest levels of conscious awareness, an open window into the subconscious and unconscious is created. This allows a clearing and integration of issues, both individual and collective, which are present within these realms.

As the subconscious and the unconscious are brought into conscious awareness and progressively cleared, the meditator is able to draw upon the contents of these memory banks, including records of simultaneous realities, multidimensional manifestations, and ultimately extending to the one, non-dual consciousness, or Reality, which lies beyond all manifestation.

THE SCIENCE OF SOUND

High-Tech Meditation uses sound as the basis for meditative focus and depends upon both modern technology and the timeless art of a Master of Meditation for the potency of its effect.

The science of sound, written about in sacred texts for thousands of years, describes pure sound as the dynamic, phenomenal, creative manifestation of Source. Sound is, in essence, Source. On the grossest level it manifests as speech, progressing vibrationally to more subtle dimensions.

"The mantra Om,
when properly intoned,
elevates unto the eternality,
the infinity of sound,
leading one from the manifest
to the unmanifest."

The power of sound to create in this sense is learned (or given) usually by study with a master of sound. The secrets of this expression of "divine" sound, which creates and sustains the universe, are held closely and in high regard by these adepts.

"The use of sound totally affects you.
Whatever sound is around you
affects you
and the effect can be observed
in your brain.
Very accelerated frequencies of sound
will affect your frequency of vibration
and produce stress.
Eventually it could make you ill.
You have to be careful what you listen to
or what you have in your environment.
But most importantly,
you can't generalize.
You can't say, 'This sound is good
for everybody and this sound is bad.'
Because human beings are vibrating
at different frequencies,
according to each one's evolution
as an energy form.
Some humans may be very accelerated
in their frequency of vibration,
and for them accelerated music is
appropriate to where they're vibrating,
and it doesn't bother them at all.
For some of them,
it may even be decelerating.
Yet for somebody who's very slow
and meditative in their vibration,
to hear rock'n'roll music is a stress.
It could make them sick.
You have to look at it
in terms of specifics."

HOLODYNAMIC SOUND

There is ample evidence that meditation with soothing music conditions the listener to be calm and relaxed. This is because such music stimulates right brain activity. In Western society, the brain is usually left hemisphere dominant. Hence, it is advisable to augment the right brain as much as possible, as such stimulation will result in increased synchrony between the two hemispheres. Such synchrony leads to a deceleration of the brain-wave frequencies resulting in expanding awareness.

The **Holodynamic** Technology of High-Tech Meditation draws on many elements of complex sound perception: rhythm, pitch, harmony, intensity, timbre, temporal order, duration, simultaneity, categorical perception, tonal memory and melody. Another important element in High-Tech Meditation is a phenomenon termed "synthetic perception". Musicians in particular frequently demonstrate this ability, which enables them to perceive intervals or harmonics as a whole. However, non-musicians, who generally perceive complex sound in terms of its component tones (otherwise known as "analytic perception") can also be trained to hear "wholes".

High-Tech Meditation utilizes the harmonic composites of varying frequencies as sound processors to entrain brain wave patterns. These harmonic composites induce a "synthetic perception" of the frequencies which are the differentials of the intervals in the harmonic composites. What is "heard" by the listener with stereo headphones are frequencies, which cannot be truly heard in the literal sense, but are perceived by the cerebral cortex as the specific intervals within the frequencies. The cortex treats these differentials as additional sounds and entrains brainwave production to them. In this understanding, the brain is being entrained or is responding much as a tuning fork will respond to another vibrating tuning fork of a similar or harmonic resonant frequency.

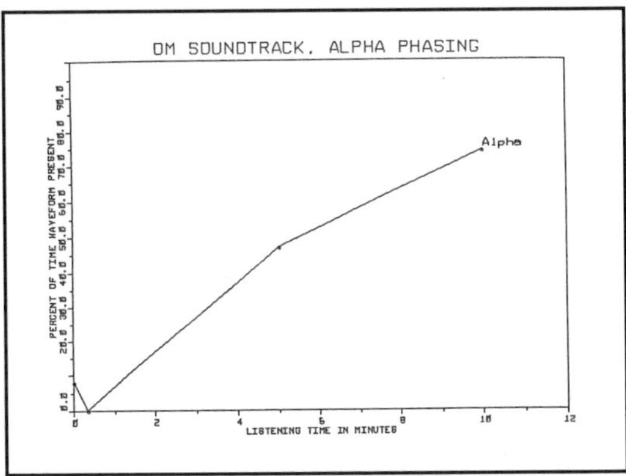

Figure 9. Graph of brain wave production using the alpha soundtrack "Om" shows remarkable increase in alpha brain wave production within a short time frame.

A balanced intensity of sound presented to the ears is especially important in High-Tech Meditation. Monaural systems are not suitable with this technology. The balance of the stereo system used needs to be carefully monitored. If the imbalance in sound intensity is too great, a situation is created in which the input to one ear is interfering with the ability of the other ear to perceive its own input. In such cases, the **Holodynamic** sound technology will be altered in its effectiveness. In practice, simple adjusting of balance between the two stereo inputs usually proves sufficient to prevent this intensity differential from occurring.

The effect of intensity levels does not mean that the ears of each High-Tech meditator must be perfectly balanced in their response to sound. However, it does mean that for each individual a balanced sound is optimal. Women, in whom musical auditory imbalance tends to be greater than in men, are advised to pay close attention to sound balancing within their stereo systems. Hearing-impaired meditators are not discouraged from using the technology of High-Tech Meditation, for the sound perceived through the vibrational structure underlying the ear itself has proved to be effective in inducing and sustaining the meditative state.

ENTRAINING WITH SUBLIMINALS

Select programs within Synchronicity High-Tech Meditation also involve the often controversial area of subliminal programming. This is a natural result of investigation into the frontier of brain function. Correctly produced subliminal messages, presented to the awareness of a relaxed and receptive individual, are extremely effective in rescripting limiting, dualistic data.

The inconsistent results achieved with subliminals stem from a lack of accurate knowledge of the mechanisms involved. The real issue is not whether subliminals work, but whether or not they are correctly scripted.

The following are requirements for an effective subliminal message:

1. It must be an affirmative statement with which you basically agree.

2. The statement must be congruent with your behavioral expressions.

3. It must be simply worded.

4. It must be repeated slowly.

5. Brain waves must be synchronized in the alpha/theta/delta range.

This last criterion, brain wave synchrony, is the key to greatest success. The effectiveness of the subliminal message is directly determined by the user's ability to be in a relaxed state of synchronous alpha, theta or delta brain wave function during its use. Otherwise, the subconscious and unconscious will not be consistently accessible and will "hear" the message only on a random basis. Either the results will be inconsistent or the message will not register at all.

While the High-Tech meditator is in a relaxed state of synchronous brain function, the subconscious and unconscious remain consistently open and receptive to whatever is presented. If it is a series of subliminal statements acknowledging truthful awareness, joy and peace, then the subconscious and unconscious will, with little resistance, accept these statements in place of old, incongruent data which limit and no longer serve one's evolutionary development.

Within the Synchronicity Paradigm, the use of subliminals is termed "rescripting" and it is not to be taken lightly, for the technique is most powerful. Most Synchronicity High-Tech Meditation soundtracks made available to the general public do not include subliminals. Those which are advertised to contain subliminals function as guided meditations which deal with such areas of general interest as health, healing, well-being and stress reduction.

In our most advanced High-Tech Meditation programs, subliminal affirmations are used for purposes of rescripting. Only those affirmations are used which promote non-dual awareness, such as: "I am Source", "I am peace, happiness and love", "I am one and free", "I am all that is". These messages are designed to clear limiting, dualistic personal data which prevent or inhibit one from achieving an experience of oneness with all that is. Such powerful non-dual subliminals, precisely programmed into the subconscious and unconscious data banks of the human brain-mind computer on a daily basis, make Synchronicity Meditation Technology the most precise and efficient transformational tool available.

Use of these advanced soundtracks requires monitoring by experienced facilitators, for the journey of one's subconscious and unconscious data is a transformation within the most elemental part of one's nature. Synchronicity, The Recognitions Program, which is our most advanced in-home High-Tech Meditation program for accelerated transformation, provides a support network to assist meditators in the integration of this impactful experience.

IS IT GOD OR IS IT ENDORPHINS?

Practitioners of High-Tech Meditation commonly experience an enhanced feeling of well-being which is, in reality, an altered state of consciousness. It differs from ordinary waking consciousness in several ways. In addition to feelings of well-being, a general sense of euphoria, and a blissful state of awareness, the meditator usually experiences a decrease in tension and stress and a quieting of the thought processes which manifest as improved clarity and focus. Transitory images, lights and colors frequently appear. This indicates the presence of a transition state in which the meditator is open to truthful insight and transcendental states of expanded awareness.

During High-Tech Meditation, the physical body often experiences a deeper state of relaxation and tranquility than is usually found in more traditional meditation. In physiological terms, such deeply pleasurable states are usually coincidental with the production of neuropeptides. These chemicals, which are produced within the cellular structure of the brain and other tissues of the human body, are associated with changes in awareness. Endorphins are one class of neuropeptide normally produced by the body which create a general sense of well-being. The discovery of this important class of naturally occurring pain-eliminators followed the observation that chemicals such as opium, heroin, morphine and codeine acted upon natural receptors located within the limbic system of the brain, and particularly within the pituitary and hypothalamus glands.

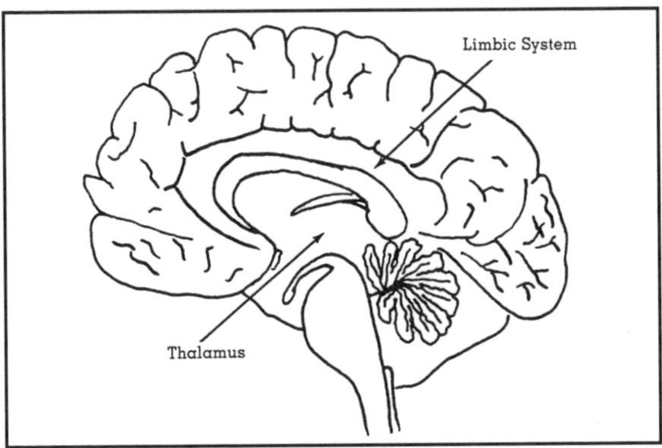

Figure 10. Cross section of the human brain reveals the location of opiate receptor sites in the limbic system and thalamus.

The limbic system has long been known to mediate emotions. Scientific research on opiates and the naturally occurring endorphins has substantiated this part of the brain as the area associated with powerful emotional response characterized by a wide range of physical somatic responses to the "released" emotions. It appears that anything which stimulates the release of endorphins will generate emotional response. Conversely, within an emotional response, we may well expect to find the release of endorphins, leading to a reduction in pain and a recovery or an increase in the feeling of well-being.

When whole-brain synchrony occurs as a result of High-Tech Meditation, the experience is one of expanded awareness, increased clarity, and greater emotional response. Improved brain wave synchrony, as observed within tracings of brain wave response measured with an EEG, indicates that the entire brain has been activated and is functioning more as a unit or a whole.

Physiological response to the stimulation of the deeper areas of the brain includes increased production within the brain of larger amounts of the neuropeptides in general, and endorphins in particular. Reactions to the stimulation provided by these "feel good" chemicals may range from an enhanced feeling of well-being to extreme states of euphoria. Hence the origin of the term "God intoxication" as applied to those mystics of yore who were well versed in the art and science of meditation. We now recognize that these individuals were actually the happy beneficiaries of hemispheric synchronization and whole brain function.

And as for the question: “Is it God or is it endorphins?”, is there really any difference? A koan for your consideration.

PART THREE:

HOW TO PRACTICE HIGH-TECH MEDITATION

WHEN AND WHERE TO MEDITATE

Meditation is best practiced in the morning, shortly after arising, when the mind and body are rested and alert. Stretching and/or breathing exercises are recommended immediately prior to meditation in order to bring the physical body to a state of more complete attention and fitness for meditation through oxygenation. Choose a comfortable area where you will not be disturbed. Sit in an upright position or in a traditional posture; do not lie down. Then, put on stereo headphones, turn on the soundtrack and listen; the technology will do the rest.

It is important that you remain alert during meditation. As the brain synchronizes and endorphins are released, the tendency is to want to lie down and go to sleep. However, the limiting data which are released from the subconscious and unconscious during meditation must be consciously "witnessed" in order to be cleared. Therefore, you need to be conscious and aware in order for rescripting to take place.

If your experience during meditation is anything but a quiet mind, do not be alarmed. A barrage of thoughts during this time is appropriate and to be expected. Just observe what is happening for you and make no judgements about what occurs. The contents of your subconscious and unconscious are simply rising to the level of conscious awareness as part of the clearing process. If you become absorbed in your thoughts, it is all right. Just notice when your witnessing awareness returns and continue watching as attentively as you can.

If you do "go out" during meditation, you may not be asleep in the way you normally understand it. Expanded states of awareness are sometimes experienced as vivid dreamlike states, often of a very pleasurable or creative nature. During such times the meditator is neither fully awake nor asleep in the conventional sense. Rather, what is experienced is a deep state of altered awareness caused by the deceleration of the brain-wave frequencies. Another manifestation of radically expanded states of awareness is the experience of heaviness or intoxication. The groggy feeling that you may have initially is simply the body's response to the presence of subtle energies released through meditation. You will become clearer as this state is integrated. Accept it as evidence that your meditation is working just as it should.

SELECTION OF SOUNDTRACKS

The depth of your meditation will be determined by the level of technology that you choose. Synchronicity soundtracks provide three levels of experience. Our wide selection of alpha range tapes (8-13 Hz.) are designed to induce relaxation and light meditation. Our theta range tapes (3.5-7 Hz.) provide a deep meditative experience and are recommended as an excellent way to begin a regular meditative practice. The delta level tapes (0.5-3.5 Hz.) are the deepest and are available only through correspondence course formats of in-home applications, such as the Synchronicity Correspondence Course and our highly regarded Recognitions Program. No one level of Synchronicity soundtrack is "better" than any other; each is a complete and unique experience.

SELECTION OF EQUIPMENT

There are a number of efficient and reliable stereo systems, including portables, that are suitable for use with Synchronicity soundtracks. An important consideration is the frequency response of the equipment and speaker systems or headphones selected. Choose those which have a good response within the 10,000-20,000 Hertz range. A stereophonic headset must be used.

The quality of the playback should be good; your sound system will need to provide you with consistent, reliable performance during frequent usage for a long period of time. Cost, surprisingly, is not the only factor in the selection of good equipment. Some of the moderately priced to inexpensive sound systems give better quality results than the more expensive makes and models. Select one within your price range that suits you and make an agreement with the dealer that you may return it if it does not perform satisfactorily over two weeks of use.

Synchronicity soundtracks are created with great care and the technology underlying the brain synchronization obtained cannot be correctly reproduced with home recording systems; therefore, reproduction of the soundtracks is not advised. Tape care is necessary to prolong the lifespan and ensure maximum efficiency of your Synchronicity soundtracks and audio equipment. Wear and tear on magnetic soundtracks can be minimized by a conscious approach to tape handling.

To extend the lifetime of both the tape and the machine, it is helpful to put the tape on a ***different*** player and fast forward and rewind your tape 5 or 6 times every few weeks to loosen the tape. This prevents the tape from becoming wound too tightly, which can lead to distortion in the sound and cause your tape player to wear out prematurely. A cleaning kit, containing cleaning and demagnetizing tapes, can be purchased at any electronics store at a cost of $5-$10. These should be used on your tape player every week or two to keep your machine in top working order.

SUPPORT SYSTEMS

Master Charles and Synchronicity Foundation are dedicated to transformational lifestyle management in all its aspects. With this in mind, we have developed specific integrative techniques designed to assist High-Tech meditators in maintaining greater expansions in their awareness on all levels - physical, mental and spiritual. These support systems will maximize your expansion and help you to integrate the elevated levels of awareness which you are accessing. This is done through a number of means including diet, exercise, breath, massage and dance.

DIET

In regard to diet, our motto is, "You eat where you vibrate." Here at the Synchronicity Foundation we recommend a vegan diet with an emphasis on live, raw foods. It is a scientific fact that all matter is energy and that energy vibrates according to the density of its form. The subtler energy forms present in live, raw foods have slower vibrational frequencies than those in heavy, cooked foods. As a result, live foods contribute to the slowing of the brainwave frequencies leading to whole brain function.

From an evolutionary viewpoint, humans are evolving from a meat-based dietary system to a plant-based dietary system, eliminating in the process animal products of all kinds. The medical ravages of diets heavily based on animal proteins are becoming well known. Not only are animal products dangerous from an overall health standpoint, but ingestion of these heavy complex proteins by meditators leads to a general densification of the form, resulting in contracted awareness.

Foods comprised of plant proteins provide higher quality nutrition and "burn clean", leaving no residue to densify the physical body and inhibit its natural functions. As meditators, we are interested in achieving maximum clarity on all levels of the human organism, from the dense physical through the subtlest levels of our multidimensional selves. From a meditative standpoint, the vibrational frequency of a pure, vegan diet will yield a more expanded awareness, not only within the physical, but also throughout our being, facilitating a greater ability to maintain clarity, focus and adaptability on all levels.

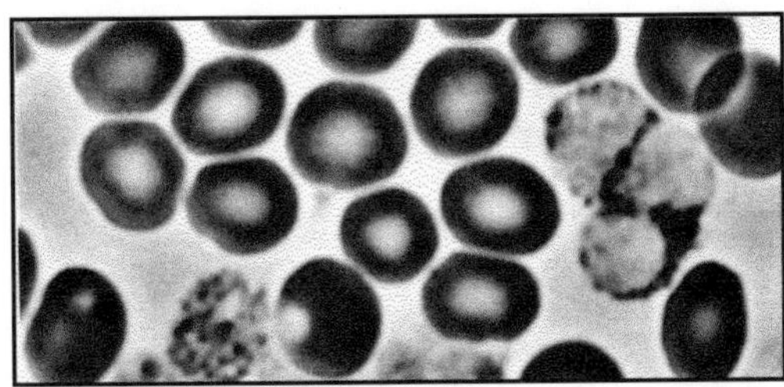

Figure 11. Live blood cell analysis reveals normal, healthy red blood cells. Larger irregularly shaped objects are white cells, which are naturally fewer in number.

This expanded awareness can be seen vibrationally as more synchronized brain wave patterns and experientially as improved functioning of the brain. By minimizing or avoiding animal products and instead relying on the much higher quality nutrition provided by plants and grains, the body becomes more refined and, as a result, the mind is able to sustain greater clarity in elevated levels of awareness. Furthermore, investigation into the benefits of raw versus cooked foods has indicated that a diet based primarily on raw foods improves the overall health of the physical body as well as the body's ability to maintain constancy in expanded states.

If you are not currently monitoring your diet to exclude heavy proteins, you should begin to do so, for the state of clarity available to you is closely related to the type of food that you include in your diet. High-Tech Meditation is the most efficient meditative discipline one can pursue and progress will be slowed unless the area of diet is addressed consciously. With experience, you will come to these conclusions yourself; they are mentioned simply as guideposts to assist you in determining an appropriate course to follow.

Figure 12. Live blood specimen reveals a condition called "Rouleaux" which fre-

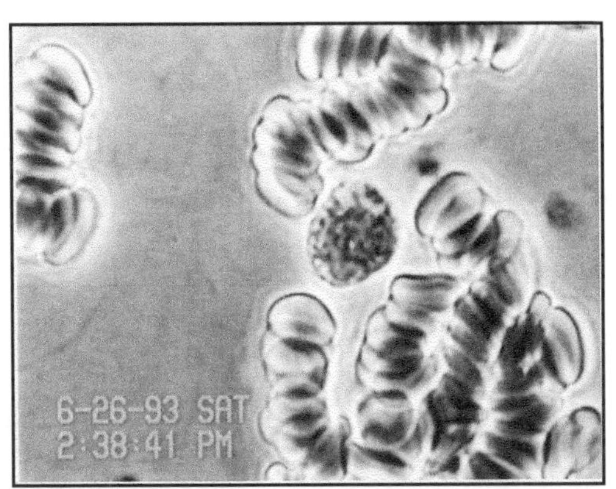

quently results from excessive protein in the body. Notice that the red blood cells are stacked upon each other, limiting their ability to transport oxygen throughout the body.

Physical detoxification is another area we emphasize, and one way to facilitate this process is through periodic fasting. For example, a one-day carrot juice fast each month is extremely beneficial and can be managed by almost anyone. Longer fasts are also appropriate and very helpful in detoxifying the body. Colonics are also recommended as a means of eliminating the toxins released by a fast. These traditional yogic practices profoundly affect your level of elevation, a fact which has been empirically validated at the Synchronicity Foundation headquarters via brain monitor and EEG evaluations.

EXERCISE/ BREATHING

It is well-known that movement and in particular the precise, balanced movement characteristic of exercise, has a beneficial effect on health and well-being. The flux of neural, hormonal and muscular energy that accompanies movement allows the physical body to remove blocks in those energy channels of the physical and subtle bodies which may be resistant to the dynamic expansion of awareness resulting from High-Tech Meditation. The itching, twitching and restlessness frequently seen in beginning High-Tech meditators are most efficiently cleared through regular exercise.

Two aspects of exercise are important: balance and breathing. Sets of exercises which are symmetrical in terms of body movements, such as walking, running or swimming, promote the balanced integration of both sides of the body. Stretching exercises, which alternate contractive and expansive movements, are also very helpful because they assist in synchronizing the sensory-motor regions of the cerebral cortex and bring the two sides of the body into harmonious cooperation. The use of stretching exercises, such as those involved in the practice of Hatha yoga, Tai Chi, Chi Kung or our contemporized version called "Synchrocize", are excellent for releasing stress and energy blockages.

Exercise assists in the integrative process of the meditative experience and also prepares the physical body for meditation in a constructive way. Oxygenation is the key word here and breathing techniques of all kinds are useful. Breathing slowly and deeply decelerates the brain waves and stimulates the production of endorphins. Including a preliminary session of breathing exercise before meditation will help you remain clear and alert as your meditation deepens.

"The breath technique was used by the Buddha unto the illumination of his soul. If it's good enough for the Buddha, it should be good enough for you."

Breathing exercises, called pranayama in Sanskrit, have long been a mainstay of meditators. Simply focusing on your breath while meditating will go a long way towards stilling your mind. It is impossible to concentrate on your breathing and to think at the same time. This is a great technique to use, not just during meditation, but any time you're feeling stressed or in need of focus.

MASSAGE

Massage, including Polarity Balancing, Shiatsu and Trigger Point Massage, is recommended as an excellent means of identifying areas of stress and tension resulting from the release of contracting, limiting data, i.e. catharsis. Frequently, as one begins a program of regular meditation, there is an initial increase in physical discomfort which can manifest in a variety of ways. Areas of chronic pain may become more prominent; there may be specific areas of tension in the body or an overall feeling of restlessness; the very act of sitting still with spine erect during meditation may at first result in stiff, sore muscles.

The bottom line in all such instances is that subtle but powerful energies are being released into a physical body which, until now, has not been required to process such energies. What is actually taking place is a progressive clearing of the physical form. Thorough attention to such areas of stress releases energy blockages which may have been present for years. The removal of such blockages not only eliminates areas of pain and discomfort, but also allows energy to flow effortlessly through the energy channels of the subtle body, opening the meditator to the experience of ever-expanding states of awareness.

In meditation programs here at the Synchronicity Sanctuary we use and recommend two types of massage. The full-body therapeutic massage, which focuses on all the major muscle groups, is very effective in identifying and relieving stress and other pre-existing conditions which primarily affect the physical body. We have also had excellent results with what we call meridian massage, which relieves and removes blockages in the subtle meridians of the body. Developed by the Chinese, meridian massage includes techniques used in Shiatsu and Trigger Point Massage. It is a very conscious and powerful way to facilitate the integration of subtle energies by focusing directly on the channels through which they flow.

Time after time, participants in our High-Tech Meditation programs have reported feelings of lightness, well-being and bliss - all characteristics of expanded awareness - which result when the physical body is appropriately and responsibly addressed. The following Brain Monitors, taken immediately before and after a 20 minute session with a professional massage therapist, graphically illustrate the clearing and synchronization which result when muscular blockages are released. We have found massage to be equally effective in releasing the body tension resulting from emotional catharsis. Such conscious attention to the meditative process results in rapid integration of the energies released through High-Tech Meditation.

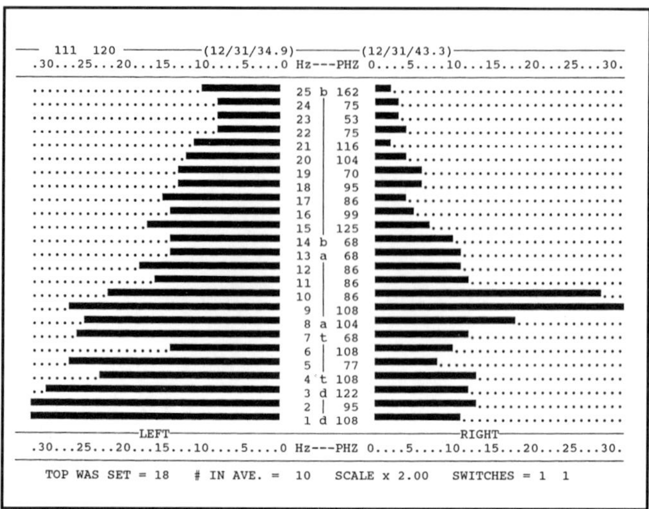

Figure 13. Prior to massage, Brain Monitor reveals body stress in the left hemisphere, indicating the presence of a physical condition on the right side of the body.

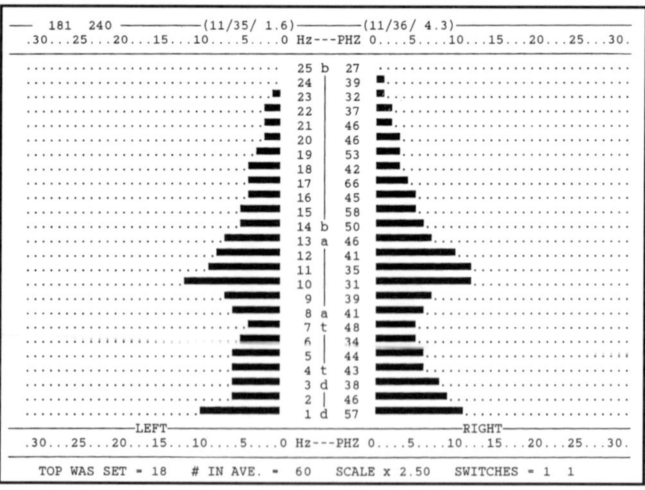

Figure 14. Following massage, the Brain Monitor reveals the equilibration which occurs when physical stress and blockages are released.

DANCE

Conscious dance is meditation in motion. Because dance has the ability to remove all manner of blocks and resistances, it is used regularly as an integrative technique at Synchronicity High-Tech Meditation retreats. We have found that dance, along with other energetic processes involving rhythm and motion, greatly assists the meditator in being free and flowing, both physically and emotionally. Individuals who may have physical limitations which prohibit other types of exercise are also able to participate to some degree in the dance. In addition to being a powerful catalyst for release, the joyful nature of dance makes it one of the best techniques for integrating and celebrating the expanded awareness that accompanies High-Tech Meditation.

And so we encourage you to dance - as integrative release, as celebration - let the human God dance! Get out of that easy chair and try it. Celebrate the cosmic dance that is happening within you. You'll be surprised how alive it can make you feel.

Great sages and their traditions from Krishna to Jesus, from the Dervishes of Sufism to the stylizations of Zen Buddhism, have utilized sacred dance to celebrate the direct experience of the human God. Hundreds of generations of spiritual seekers can't be wrong.

PART FOUR:

WHAT TO EXPECT FROM HIGH-TECH MEDITATION

CATHARSIS AND INTEGRATION

A variety of phenomena may occur during the practice of High-Tech Meditation. These phenomena are identical to those which arise in any meditative practice. The precision of Synchronicity High-Tech Meditation will, however, accelerate their appearance. The range of expression of these phenomena is virtually unlimited. One is advised simply to observe and recognize them as temporary manifestations along the way to more integrated transcendental experience.

Initially, the High-Tech meditator can expect a generally calm, peaceful and deeply relaxing experience. Insights or flashbacks to earlier life experiences may occur. An awakening or expansion of the natural energy dynamic within the body will also begin. It may take a variety of forms but is usually experienced as the calm and peaceful sensation mentioned above. Other forms may include periods of grogginess, heat, tingling, minor itches and twitches, and/or spontaneous body movement.

As the practice of meditation continues, old issues that have remained unaddressed will surface. Insights occur which reveal how suppressed experiences create limitation. The emotions accompanying these experiences may find expression in episodes of anger, fear or sadness, alternating with periods of joyful understanding and expanded awareness.

The surfacing of old, outdated subconscious and unconscious data into the conscious mind is termed "catharsis". There are three main areas in which High-Tech Meditation initially catalyzes catharsis:

1. For most people there are physical issues that need to be clarified and addressed. Old illness patterns may reappear as physical blocks which need to be addressed in order to further expansion. Issues concerning diet, health and body awareness are likely to be addressed and clarified in light of the meditator's expanding awareness. Any substance abuse or patterns of self-neglect will also present themselves as incompatible with the meditative lifestyle.

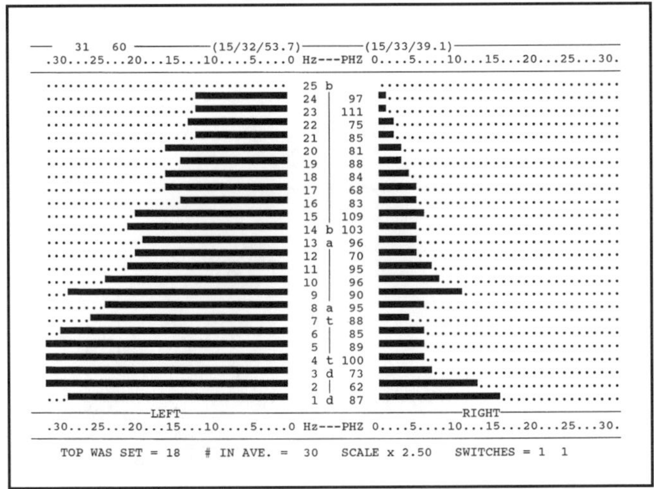

Figure 15. A Brain Monitor reading of this type indicates physical catharsis, which results when energy is not able to flow freely through the body.

2. High-Tech Meditation stimulates the emotions in a subtle but most effective fashion. As a result, patterns of emotional expression and/or repression become apparent as you begin the meditative journey. If there are emotional issues that are not being clearly addressed, you will most assuredly get in touch with these during meditation, so be alert.

3. Finally, you may find that mental activity increases. Racing minds during meditation are not uncommon and, in some cases, the thinking process appears intensified.

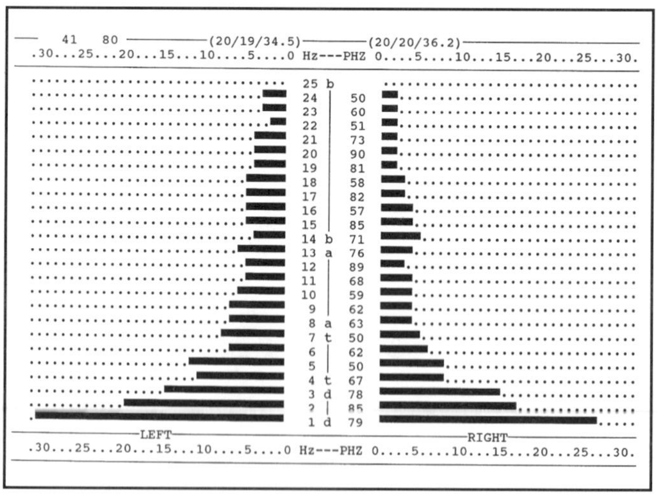

Figure 16. Catharsis is measured by the numbers in the right-hand column. Higher numbers indicate a greater degree of asynchrony, or catharsis. Zero represents perfect synchrony.

These periods of 'catharsis' are cyclical and are followed by periods of integration. Upon integration of the new insights developed through the use of Synchronicity High-Tech Meditation, the meditator's life begins to transform, sometimes radically. Relationships may undergo revamping or decisive renewal; occupations may shift; lifestyles may change dramatically.

Admittedly, this sounds like a lot to handle, especially when what you are seeking is a little peace and quiet. However, meditation is perhaps best understood as a clearing process which progressively reveals a more truthful perception of Reality.

Master Charles often refers to meditation as "a confrontational insanity" and "an upheaval in one's being". Viewed from this perspective, catharsis is recognized for what it is - a highly beneficial part of the cyclic evolutionary process whereby the subconscious and unconscious are cleared of fraudulent, limiting data. Such a clearing is essential before truthful perception can occur. Be assured that the sometimes uncomfortable process of cathartic release is followed by deeply experienced feelings of joy and well-being. The insights resulting from such precision meditation are unlike anything else that you can experience.

In the classical meditative systems, would-be meditators underwent years of purification and concentration exercises before they were initiated into meditation. Typically, students were placed in appropriately isolated locales where they could avoid distractions, save for those created by their own minds. Given enough years of practice, seekers were able to achieve a state of mental equilibrium with the resultant expansions of awareness.

Traditional purificatory processes cleared much of the dualistic data, resulting in less catharsis for the meditator. However, Westerners are very naive about the process of catharsis and sit for meditation without preparation. As a result, meditation brings forth all their dualistic, limiting data. Again, it must be emphasized that the experience of catharsis is essential to the process of purification as dualistic data is released and a non-dual rescripting takes place. True meditation is that which brings forth the greatest upheaval and in its integration leaves the greatest clarity.

"The experience of expanded awareness overpowers the experience of limitation."

This mystical view of purification is validated by the new science. In psychological terms, meditation forces us to confront the subconscious/unconscious data stored in the human brain-mind computer. As a result, old information is brought into conscious awareness, providing a glimpse of all the conflicting data. If the data is very dualistic the disturbance will be maximum, less so if it is non-dual. Limiting data will be more quickly and easily discarded if one is able to bring a measure of detached awareness to the observation of its release.

"You create your stress through the non-allowing of what is happening and thereafter make a judgemental interpretation of it."

It is most important that the meditator consciously address this process of release since the retention of unconscious blockages prevents the expansion of awareness. One must clearly understand that clarity and awareness cannot increase until fraudulent, dualistic data are released and replaced by truthful, non-dualistic data.

In the Synchronicity Recognitions Program, which is our "industrial strength" delta level in-home High-Tech Meditation program, we require regular telephone contact with participants so that we can monitor any catharsis that arises and recommend appropriate integrative techniques.

As you progress in the meditative experience, the cycles of peak, breakthrough,catharsis, integration and illumination still occur. It becomes obvious, sometimes very quickly, that meditation does indeed work as periods of peace, bliss and relaxation increase in frequency and duration. While there may still be intense cathartic episodes, the meditator invariably finds that life is increasingly experienced as a non-serious play of consciousness.

ACCELERATED TRANSFORMATION

After years of research into the brain wave patterns of regular users of Synchronicity High-Tech Meditation technology, our researchers produced the following study comparing the effects of High-Tech Meditation with traditional meditative systems.

A Comparison Study of Synchronicity High-Tech Meditators With Zen Monk Studies

Introduction

One of the principal differences between Master Charles' Synchronicity Holodynamic Technology (High-Tech Meditation) and traditional orthodox (low-tech) systems of meditation is the speed with which measurable changes occur in one's states of awareness. This has been confirmed by measurement of the brain wave patterns produced by regular users of Synchronicity Meditative Technology over a seven-year period. These were then compared to the measurements produced by those using traditional meditative systems.

There are a number of documented cases where groups of meditators using classical methods of meditation were measured. The most notable are the measurements made on the Dalai Lama's attendants (roughly 20 monks) during the early 70's, and, more recently, those reported by Tomio Hirai in the book Zen Meditation and Psychotherapy after measuring 40 Zen monks.

It should be noted that these studies represent the very best results that can be expected from the classical meditative disciplines as these individuals were totally immersed in the contemplative lifestyle. Even those who are resident at the Synchronicity Sanctuary do not have the fully structured meditative lifestyle of cloistered monks. Further, most participants in the Synchronicity Recognitions Program (the correspondence format program which utilizes Synchronicity High-Tech Meditation on a daily basis) live everyday lives in cities, working normal jobs, raising families, and enduring the distractions which those experiences engender. Not at all what one would define as a contemplative lifestyle.

The results of the two studies of classical meditators were basically the same. Novices (their term) were considered to be those who had five years or less meditative experience. These individuals produced mid to high frequency alpha waves (10 to 12 Hz.). Moderately experienced meditators (10 to 20 years experience) continually produced low frequency alpha (7 to 9 Hz.), the actual frequency being lower as experience increased. Very experienced meditators, those with 20 to 40 years experience, consistently produced theta frequencies in the 5 to 6 Hz. range. An interesting aside is that the experimenters made no mention of the other bands of frequencies. Delta (0.5 to 3.5 Hz.), has long been considered to be the most notable sign of an expanded state of awareness typical of enlightening beings. Alpha (7 to 13 Hz.) is considered necessary for attentive focus, and beta (14 to 25 Hz.) is necessary to retain conscious awareness. In the absence of beta and alpha, the presence of delta indicates Stage II or III sleep.

The reason the other frequencies are not mentioned in these studies is probably due to the fact that the researchers used traditional electroencephalographs (EEGs) for their measurements. The wave forms from EEGs are characterized by being mostly alpha, or primarily theta, or REM, or sleep spindles, etc. This is because the data from EEGs are complex superpositions of frequencies which are difficult to decompose into more fundamental constituents. Thus, they cannot say much about specific frequencies, other than the dominant ones.

Similar research by C.M. Cade and N. Coxhead (The Awakened Mind, Element Books Ltd., Longmead, Shaftesbury, Dorset, England, 1989) involved an instrument called "The Mind Mirror", which used analog filters to transform the data to the frequency domain so that brain wave frequencies could be observed instead of the trains of composite amplitudes characteristic of the EEG. One of the results of their measurements was a delineation of the profile of frequencies present for various states of awareness. Their state V, which they call the "awakened mind", reflecting the state of enlightening awareness, is comprised of beta (typically from 15 to 17 Hz.), along with a noticeable peak in alpha (in the 7/8 Hz. region), a theta bulge (in the 5/6 Hz. area), and expanding delta (i.e. increasing in amplitude as the frequency approaches zero). Many Synchronicity Recognitions Program meditators exhibit this state V response on the Brain Monitor, which also displays data in the frequency amplitude domain.

RESULTS

Our extensive research has shown that as they begin the Recognitions Program, most people produce alpha at 11 or 12 Hz. This is consistent with patterns demonstrated by novice Buddhist monks. As individuals move through the Recognitions Program, however, their brain wave patterns shift, the alpha peaks becoming lower in frequency, larger in amplitude, and more persistent in time. As they acquire more experience, we see more theta and developing delta. Symmetry improves throughout. The left to right coherence which we call synchrony factor often varies as a function of the moment. It reflects the bliss as well as the catharsis often experienced by a meditator and is a measure of the equilibration experienced at the time of the measurement.

The accompanying Brain Monitor results are typical of our findings. These scans include both residents of the Synchronicity Sanctuary and non-resident participants in the Recognitions Program. The major point of interest is that none of these meditators has more than 7 years experience with the Synchronicity Holodynamic Technology - most of them less than that - yet they demonstrated brain wave patterns consistent with advanced meditators (20 years or more experience) in the Buddhist monk studies.

These findings empirically validate the assertion first advanced by Master Charles in the Synchronicity Paradigm in 1983, that Synchronicity Technology yields a fourfold acceleration factor over classical methods of meditation. The following data are presented for purposes of comparison of Synchronicity meditators with the Buddhist monk studies in terms of measured dominant brain-wave frequencies in relation to years of meditative experience.

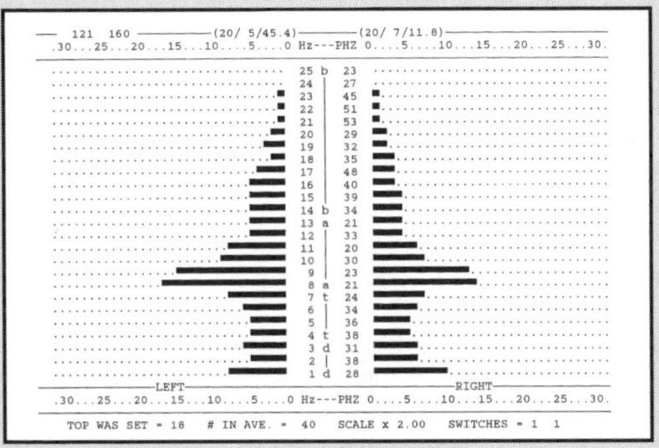

Figure 17. Subject 1: 5 years - prominent 8 Hz. line - theta/delta at 3 Hz. - developing delta.

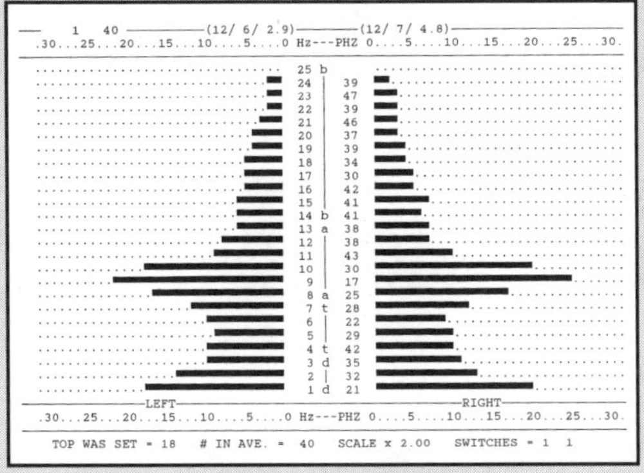

Figure 18. Subject 2: 6 years - strong alpha at 9 Hz. - theta at 4/5 Hz. - good delta production.

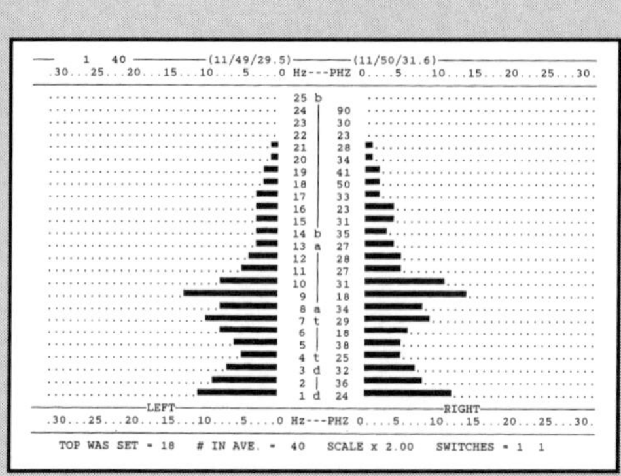

Figure 19. Subject 3: 6 years - alpha lines split at 9 and 7 Hz., good symmetry - delta developing.

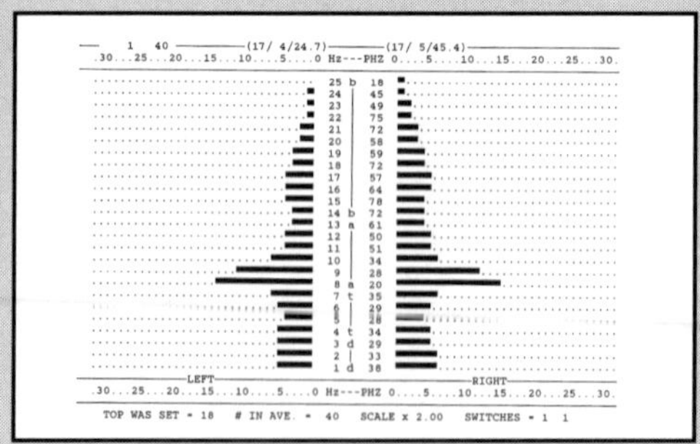

Figure 20. Subject 4: 3 years - 8 Hz.alpha, good symmetry.

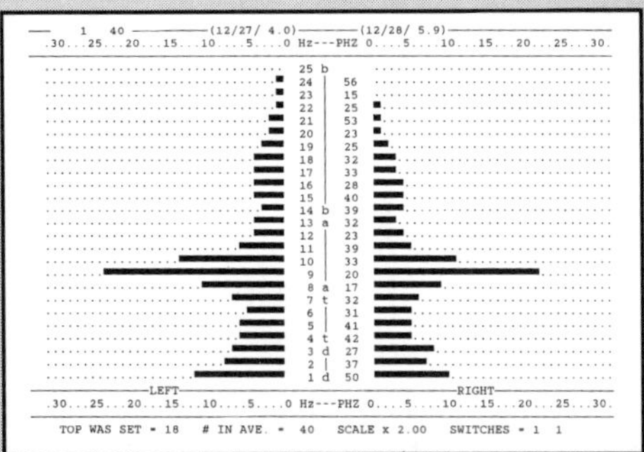

Figure 21. Subject 5: 4 yrs. - strong 9 Hz. alpha - theta/delta at 3 Hz. - delta developing.

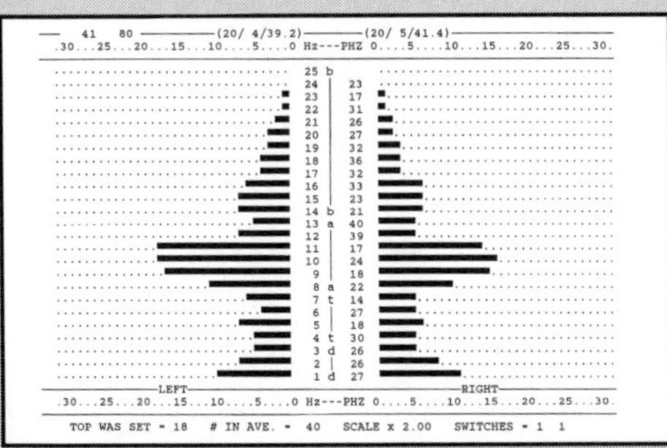

Figure 22. Subject 6: 7 yrs. -alpha band 8/9 to 11 Hz. - clean theta line at 5 Hz. - delta developing.

References

Tomio Hirai, Zen Meditation and Psychotherapy, Japan Publications, Inc., Tokyo and New York, 1989.

Kasamatsu, A. and Hirai, T. "Science of Zazen", Psychologia, 6, 1963, pp. 86-91.

THE EXPERIENCE OF AWAKENING

As mentioned earlier in this section, there is an awakening of the natural energy dynamic within the body when you begin High-Tech Meditation. Such awakenings have traditionally been regarded as milestones on the human evolutionary journey and are characterized by the release of energy throughout the neurosensory systems of the body.

Synchronicity High-Tech Meditation is based on the understanding of the inherent balance or synchronization within the brain which is required to produce the awakening experience. When the awakening occurs, the natural energy residing in the human body moves in a spiraling motion up the spinal column. During meditation, when the body is still and the mind relatively quiet and introspective, this spiraling energy may become noticeable as it is permitted freedom of expression. This often causes a meditator's body to spontaneously rotate and move in a circular, spiraling form. Other physical indications of such an awakening include feelings of heat or the movement of energy up the spine. Breathing patterns may alter and spontaneous retention of the breath may occur. In certain individuals, this energy may produce other involuntary patterns of movement. Such movements are well documented in the meditative literature of the East, and many yogic postures and techniques originated as attempts to duplicate these spontaneously produced movements. Occasionally a latent natural energy release occurs suddenly and explosively. The onslaught of so much available energy causes a wide range of sensory, hormonal and muscular phenomena within the meditator's body. On subtler mental and emotional levels, purificatory release may also occur, such as spontaneous laughing or crying.

The purpose of all these movements is to release blockages and to clear the body and nervous system so that the meditator can endure the energy of more expanded states of awareness. This energy is widely recognized in the various Eastern philosophical systems, where it is known as kundalini or chi. It is the same energy which leads to the spiraling motion and all of the other awakening processes commonly seen in High-Tech meditators. Though it lies latent at the base of the spine of every human being, the vast majority of people live out their lives without ever knowing it's there.

In the Eastern traditions, this energy is understood to be the primordial or cosmic phenomenal energy which awakens as a result of deep meditation or contact with a true master. Because both of these conditions are met in the case of Synchronicity High-Tech Meditation, the awakening process does occur. When awakened, this energy begins to move upward through the subtle central channel of the spine, piercing the energy vortices of the subtle body and initiating various processes which bring about the total clearing of the entire being.

While Western science has little awareness of the awakening experience, its existence is well-known among Eastern traditions. Such awakening is a normal and natural evolutionary development in a meditator and is not something to be feared. It arises spontaneously when the time is appropriate, and one should not attempt to force it into premature arousal. The guidance of an enlightening master is required in this area as one begins to access the realm of these subtle but very powerful energies.

ENLIGHTENING HARMONICS

A "multidimensional" understanding of the human form and how it manifests is necessary in order to Sourcefully master the physical vehicle in which we find ourselves. According to both contemporary science and

Eastern philosophical systems, everything is vibrating energy. The sacred texts of the Eastern traditions go on to describe the actual process of physical manifestation, i.e. creation, through which energy becomes form. The energy dynamic of phenomenal existence, which at the most fundamental level is vibrating at the frequency of light, must reduce or contract its frequency of vibration in order to manifest physical form. The term "transduction" is used by Master Charles to describe this process of reduction of the frequency of vibration.

Beginning with the supracausal level of experience, at which stage the soul is in essence an undifferentiated energy vibrating at the frequency of light, the energy is transduced, or densified. This slows its frequency of vibration, which provides a subtle vehicle for the soul essence called the causal body. On the causal level of experience, the data bank corresponds to the unconscious mind, with its primal, instinctual data relating to the initial dualistic polarization within energy, the subjective/objective, the masculine/feminine, the noumenal/phenomenal, etc.

When this energy is further contracted, or densified, it creates what is known as the subtle body, which corresponds to the subconscious data banks. These relate more to the specific, personal data of the individual. The final level of contraction of energy results in the formation of the dense physical body. Thus, four levels of transduction are required to produce the human form. Western psychology currently recognizes the three lower levels, the conscious, subconscious and unconscious and has recently begun to embrace the Eastern understanding, which includes an awareness of the supracausal level of human experience, also known as witness consciousness.

The individual personality essence, or what is termed the soul, is evolved within the three higher bodies, the supracausal, the causal and the subtle. These three bodies operate as a unit and are not dependent upon the physical body. When the soulular essence departs the human form, the physical body dies and the other three continue to exist, carrying with them the memory of all the individual experience from that and all simultaneous realities.

This soulular memory remains, for the most part, unconscious. However, this information is available and can be retrieved. The accessing of these dimensions, which is in essence the constant experience of supracausal awareness through the physical form, is what is known as enlightenment.

The process through which this radically expanded state of awareness is achieved is essentially one of harmonic access. There is a harmonic relationship which exists between all four bodies, such that if you hit the musical C note in one body, it simultaneously vibrates the other three bodies with the same C, but at higher octaves. When there is alignment of these frequencies, you have the truthful perception of the Sourceful, soulular, true Reality essence harmonically reverberating through all the bodies.

A certain number of frequencies, corresponding to the supracausal, causal and subtle bodies, exist within the physical form which are harmonically linked with the three subtler bodies. However, it is very rare for such harmonic reverberation to occur. There are so many accelerated frequencies playing through the physical form that the harmonic "window of opportunity", so to speak, is missed. The harmonic window for the physical body is a very slow, decelerated frequency. Unless the no-mind state is achieved and the entire physical apparatus is entrained to vibrate at these low frequencies, there is no accessing the harmonic window which opens to higher awareness. Until this occurs, the truthful perception of yourself as Source is not possible.

Detailed knowledge of enlightening harmonics and the specific frequencies involved is limited to those masters of sound whose multidimensional awareness allows them access to the subtlest dimensions of physical experience. In the West the science of sound is still in its infancy and for the most part remains an extremely esoteric subject. Yet thanks to a few visionary individuals, the creative powers of sound are being harnessed and made available on an unprecedented scale for the benefit of humankind.

Synchronicity High-Tech Meditation is one such example. Synchronicity Technology stands alone among brain-entrainment devices because it decelerates and entrains the specific frequencies required to provide access to the harmonic window of the physical dimension. Its precision facilitates the transduction process, providing access to the subtlest multidimensional levels of supracausal Source Consciousness Awareness.

"I am the forever aware witness as one Source consciousness.....
this is the eternal reverberation of all and everything."

Even a rudimentary understanding of enlightening harmonics makes it apparent that High-Tech Meditation is an extremely advanced and powerful scientific technique for the transformation of human awareness. However, there is another very important catalyst which must be present for full illumination to occur. This is the subject of our final chapter.

PART FIVE:

MASTERSHIP: A CONTEMPORARY UNDERSTANDING

THE PRINCIPLE OF MASTERSHIP

In all traditions in which meditation is the dominant technique in the journey to a constant awareness of the enlightening state, the art and science of meditation has been revealed through the auspices of a teacher or enlightening master. Those who genuinely wish to access higher dimensions of awareness must always keep in mind that mastery of the human form, which includes the mind, must be pursued in cooperation with an enlightening teacher. The states and processes through which one must pass require the energy dynamic of a master who can assist in the opening of specific energy vortices, known as chakras in the Eastern traditions, which are present in the subtle bodies of the human form. It is not possible to open these vortices and release the blockages therein on one's own. The energy, knowledge and experience required to do so far exceed the abilities of even the most ambitious seeker. Enlightening masters create methods of meditation that are effective in bringing their students into a state of awareness from which further development proceeds. Synchronicity soundtracks are unique among meditation tapes in that they were developed for contemporary meditators by a Western enlightening master, Master Charles.

The enlightening energy dynamic of mastership is a subject not easily understood in the West. The East, however, has a long tradition of mastership as the means of attaining constancy in expanded states of awareness. Since the West has no such tradition, our culture remains, for the most part, "guru-phobic" and skeptical of the principle of mastership. Until you directly experience the energy dynamic of an enlightening being, the entire concept may well seem implausible. Nevertheless, if you truly wish to follow the meditative path, the assistance of a master is required.

"The psychology of the West is limited because it is modeled on the non-aware."

In terms of the evolution of its collective soul, Western consciousness is still in the adolescent stage. Up until now, those relatively few mature souls who recognized the significance of mastership sought out their teachers in the mystery schools of the East. However, this is rapidly changing. With the advent of the New Age, we are witnessing large numbers of Westerners who are demanding expansion and elevation, and from an evolutionary standpoint, it is inevitable that their demand will be met. In fact, this has already begun and it is most interesting to note that it comes in a form which is appropriate for the technically advanced Western, left-brain dominant mind.

"We Westerners have created fast food. Why not fast enlightenment?"

One facet of modern technology which has proven to be a boon to High-Tech meditators is the fact that the energy dynamic of the master can be transmitted onto the magnetic tape used in the production of audio and video cassettes through sound vibration. The elevating and enlightening energy dynamic is actually contained within the sound, enabling the listener to directly experience the benefits of the master's expansive reverberation.

In much the same way that psychics and psychometrists are able to scan the biofield and obtain information, the master's energy field, by virtue of its ability to entrain magnetically, modifies the energy of the tapes so that the enlightening presence is holographically encoded into their structure. For this reason alone, meditation tapes that carry such an energy dynamic are in a class by themselves. Remember, a magnetically recorded soundtrack carries the frequency of the one who makes it, and this becomes a part of the spectrum of frequencies that the user absorbs.

When considering soundtracks for use in meditation, one must be aware of how they were produced. Soundtracks which simply contain relaxing music or purely technologically oriented methods of synchronizing the brain's hemispheres will never be able to expand the awareness and hold its focus in the entraining energy of enlightenment as Synchronicity High-Tech Meditation soundtracks are able to do. The simple truth is that the unenlightened can never entrain enlightening experience. When this is clearly understood, the use of soundtracks created from a state of enlightenment to entrain a state of enlightenment becomes the obvious choice for the attentive meditator.

Inevitably, when the demand for mastership arises, there will always be those non-illumed individuals who declare themselves masters for their own capitalistic purposes. They are quite easily recognized by the mere fact that nothing Sourceful happens in their presence. The same might be said of mind toys and brain machines which offer promises of "instant meditation" or a "supercharged brain". Given today's technology, it is actually quite easy to entrain the brain's right and left hemispheres into various states of synchronization or balance. However, 20 minutes later when the tape is turned off and the initial endorphin stimulation diminishes, the brain returns to its previous level of imbalance. Should negative catharsis arise, the meditator blames the tape and stops using it, thus sabotaging the meditative process.

The addressing of human catharsis is an essential part of the meditative experience, yet it remains largely unacknowledged and unaddressed in the West. Those who market soundtracks and brain machines simply as a capitalistic business venture know little or nothing about meditation or what is actually going on in the brain. They do not have the full enlightening experience and training in mastership, which is the means classical systems have traditionally employed in the human transformational journey known as meditation.

The true master, on the other hand, continually catalyzes meditation through his energy dynamic. The master's very presence is the epitome of "high" technology. He exists to entrain, elevate and illume - nothing more. Thus, the one and only focus of Synchronicity Foundation is the art and science of meditation. For the past 10 years, under the direction of Master Charles, we have assisted thousands of people in the experience and integration of radically expanded states of meditative awareness. It is all we do.

And so we offer this word of caution: pay close attention to the source of any brain entrainment technique you are considering. Without mastery as the underlying principle, no technique, discipline, methodology or technology will provide responsible, consistent meditative unfoldment.

"The unenlightened will never produce enlightening technology. Einstein didn't produce his famous theory from an experiential background in cooking."

To understand more clearly the need for a master, one must consider the mechanics of the process of enlightenment. A simple analogy is that of a tuning fork. The nature of the fork is to mimic whatever vibration is in its range of experience. Because the master is constantly vibrating with such intensity at a slower, more decelerated frequency, his vibration tends to attract and entrain other frequencies to its own level of deceleration. In the master-disciple relationship, the master's vibration is so powerful that it entrains the disciple to vibrate at lower frequencies than normal, thus opening the disciple to new dimensions of experience. As a result, a sympathetic harmony is achieved between the master and disciple. When considered from this point of view, the mysterious state of illumination is simply the complete and constant deceleration of the brain's frequency into the same state of non-dual awareness as that of the master.

"Mastership is the lazy man's way to enlightenment. It's like the sun for one who wants a tan. You must simply sit in front of it. It's perfect for Westerners."

Thus, while contemporary technology can assist one in accessing non-dual states of awareness, the assistance of an enlightening master is required for successful integration and continued expansion. Entraining and catalyzing the experience of Source for students is the function of a master. Authentic masters of meditation take this responsibility to heart and live it.

"I am what I am
because I have in every moment,
with every breath, demanded
more expanded awareness.
So if you would wish the same
to manifest in you,
let nothing come between you
and your God."

EPILOGUE

For over 10 years, Master Charles and Synchronicity Foundation have been the leaders in the exploration and refinement of the High-Tech Meditation Experience. At the present time, Synchronicity Meditative Technology provides the most precise and efficient means available for creating constancy in human equilibration, thus making the enlightening experience available on a scale never before known in the history of man.

Having achieved unparalleled success in the development of technology for awareness expansion, we have widened our focus to include the previously unresearched area of High-Tech integration.

Based on our findings, Synchronicity Foundation has developed specialized Integrative Protocols which radically accelerate the progressive release and integration of catharsis, thereby further contemporizing the age-old process of purification which is an essential part of the meditator's journey.

As a result of our empirical research, not only can we chart your meditative process over days, weeks, months and years in relation to your brain function, but we can also identify blockages - mental, emotional and physical - as they appear within the subtle energy fields. We can determine the exact location of these energetic blockages, and can help identify the repetitive patterns in behavior and thought which result in limitation and contraction in awareness.

Our one-of-a-kind computerized "Brain Monitor", developed by our own empirical research team, reveals the complete brain function picture produced before, during and after the use of Synchronicity High-Tech Meditation. Brain monitoring can be done on an ongoing basis in order to track the meditator's progress over time.

Our Biofeedback Rescripting research has demonstrated precision reprogramming of limiting, dualistic data. Using this technique, High-Tech meditators learn to consciously access the windows of the subconscious and unconscious in order to rescript self-empowering, non-dual data. Individualized programs are designed which enable meditators to identify, address and release their own limitations, resulting in ever greater meditative expansions in awareness.

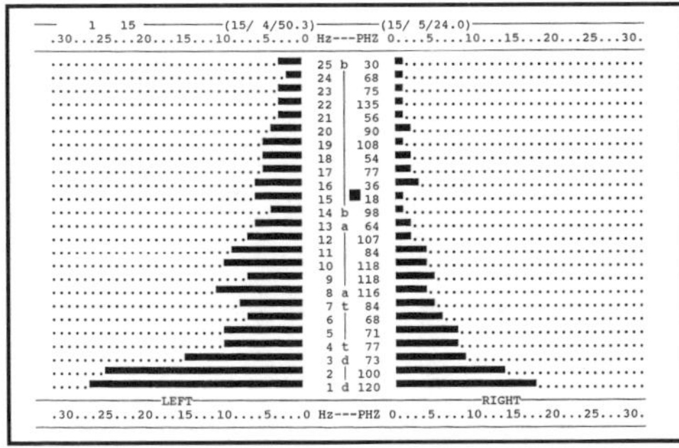

Figure 23. Brain Monitor taken prior to theta window biofeedback training.

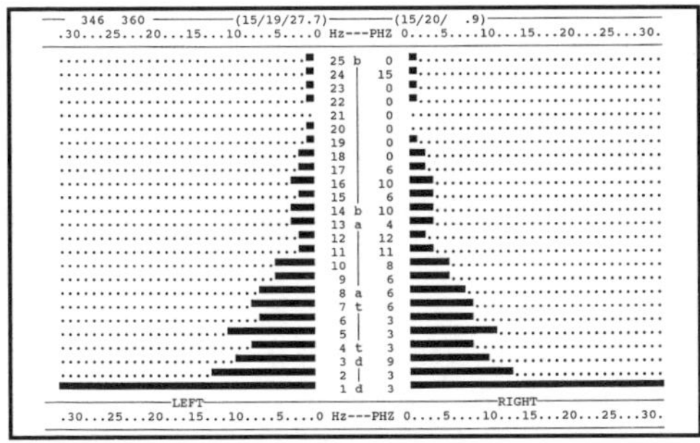

Figure 24. Brain Monitor taken following biofeedback training reveals improved hemispheric synchrony and a significant increase in theta production.

Comprehensive training and direct experience of Synchronicity technology and associated integrative techniques are available through the Synchronicity School of High-Tech Meditation under the exclusive guidance of Master Charles. The school provides full diagnostic, integrative and counseling services as well as daily programs with Master Charles. Within this context of greatly expanded awareness, participants experience radically accelerated transformational growth.

We are fortunate to live in an age when the enlightened use of contemporary technology allows humanity to dramatically accelerate its evolution. Collectively we have developed the technology which permits such advancement. Our collective desire has manifested a contemporary master to demonstrate its efficiency and precision. And together we shall move forward to a new, more illumed state of human experience. If you find yourself interested in Synchronicity High-Tech Meditation, you have come within the realm of an accomplished master whose heart is focused on seeing your meditation bear fruit.

"Remember that I live
to be fully alive
and to share such awareness
with you."

Having manifested such a master indicates that you have arrived at a major crossroads in the human journey and your choice at this moment will affect the course of your evolutionary development. With this in mind, we wish you well on your journey into the world of High-Tech Meditation, the fast track to enlightenment. No matter where your journey takes you, be assured that, in truth, you exist in a state of Sourceful celebration and in this you are eternally appropriate.

May you enjoy the ecstasy of peace.

SYNCHRONICITY FOUNDATION

Synchronicity Foundation is a not-for-profit, non-denominational, non-sectarian organization dedicated solely to the expansion of Source Consciousness Awareness. To this end, the Foundation offers a wide variety of products, programs and services designed to unfold the enlightening experience within each individual.

Since its inception in 1983, Master Charles and Synchronicity have achieved international status as leaders in the field of High-Tech Meditation. For over 10 years, the Synchronicity Foundation has diligently researched and documented the High-Tech Meditation experience in hundreds of thousands of daily usages. At the Synchronicity Retreat and Research Center in Virginia we offer the most comprehensive experiential and integrative assistance available for the contemporary meditator.

THE SYNCHRONICITY SANCTUARY

Synchronicity Foundation is headquartered on a scenic Retreat and Research Center located in the Blue Ridge Mountains of Central Virginia. This intensively meditative environment, which revolves around Master Charles and his exceptional state of enlightening awareness, is home to the Synchronicity Resident Community, a group of individuals committed to maximum transformation and acceleration of the human evolutionary journey through the practice of Synchronicity High-Tech Meditation.

The ongoing study of the transformational lifestyle exemplified by the Synchronicity Resident Community offers clear insight into the possibility of contemporary enlightened living, regardless of one's situation.

REFERENCES

1.Daniel Goleman. 1988. The Meditative Mind. New York: Putnam Publishing Group.

2. Master Charles. Nov.11, 1984. "The Ecstasy of Peace: A Revolution in Consciousness". Lecture at Barbizon Plaza Hotel, NYC.

3. Wallace B. Mendelson, J. Christian Gillin, and Richard Jed Wyatt. 1977. Human Sleep and Its Disorders. New York: Plenum Press.

4. Michael Hutchison. 1986. Mega Brain. New York: Ballantine Books.

5. Ilya Prigogine and Isabelle Stengers. 1984. Order out of Chaos: Man's New Dialogue with Nature. New York: Bantam Books, P. XXI to XXII.

6. Patrick Harbula. 1987. "Sounds of Transformation: A Talk with Master Charles". Meditation Magazine. Vol.2, Number 4, P.20 to 28.

7. Doreen Kimura. 1973. "The Asymmetry of the Human Brain." Scientific American, 228(3), P.70 to 78. (See also: D. Kimura. 1967. "Functional Asymmetry of the Brain in Dichotic Listening." Cortex, 3(2), P. 163 to 178.)

ABOUT THE AUTHORS

Cynthia Larsen has been a student of meditation and Eastern philosophy for 15 years. She attended the Universities of Minnesota and Wisconsin and California State University, majoring in Psychology and Religious Studies. She worked as a writer and a counselor before joining the staff of Synchronicity Foundation.

Paul Shannon is a physicist who has worked both in the government and industry in pure science as well as in applied science and development. He is the originator of the Brain Monitor, which is used in Synchronicity Foundation research and programs. He currently advises the Foundation in technical areas and on matters regarding research.

Index

A

acceleration 38, 82
affirmations 53
alpha 40, 61, 84
alpha brain waves 34
analytic perception 48
awakening 90, 92

B

behavioral modification 11
beta 40, 84
beta state 26
blissful intoxication 29
blockages 81, 91
brain hemispheres 24
Brain Monitor 77, 86, 111
brain wave 49, 86
brain wave patterns 86
brain wave production 41
brain wave synchrony 52
brain waves 40, 42, 69
brain-wave frequencies 38, 48
breakthrough 44, 82
breathing 69

C

catalyst 97, 106
catharsis 44, 70, 76, 77, 78, 79, 80, 82, 103, 110
cerebral cortex 49, 68
chi 91
concentration 19
consciousness 17, 38

D

dance 73
data 37, 39, 52, 76, 81, 93, 111
data banks 26, 37, 44
data processing 38
deceleration 105
delta 61, 84
delta brain waves 38
detachment 19, 81
detoxification 67
diet 65, 77
doorway 37
dreamlike states 60
duality 17

E

EEG 41, 43, 56, 67
electromagnetic energy fields 25
empirical 23, 29, 86, 110
endorphins 29, 54, 56, 57, 60, 69, 103
energy 22, 91, 93, 94
energy blockages 70, 110
energy bodies 94
energy dynamic 99, 100
energy fields 102, 110
energy vortices 92, 99
enlightening dynamic 20, 31, 84, 96, 100, 102, 106
enlightenment 101, 113
entrainment 22, 39, 51, 96, 106
equilibration 17, 24
euphoria 54, 57
exercise 68

F

fasting 67
food 65

G

"God intoxication" 57

H

harmonics 92, 95, 96

hemispheric synchrony 31, 42, 57
"high" of life 29
High-Tech Meditation 20, 30, 50
76, 78, 92
Holodynamic Technology 11, 48
50, 83
holographics 102
human journey 15, 29, 90,
104, 113

I

illumination 44, 82, 97, 105
integration 44, 45, 61, 69, 71,
73, 79, 82, 109, 110
integrative techniques 64

K

kundalini 91

L

lateralization 24
left-brain 21, 26
left-brain dominance 101
limbic system 56

M

massage 70
master 7, 46, 92, 99, 106
Master Charles 10, 11, 13,
44, 64, 83, 86, 93, 100, 112
mastership 7, 13, 99, 100
mastery 104
meditation 17
meditator's journey 110
memory 94
meridian massage 71
mind 17
mind toys 13
mindfulness 19
monks 83, 86
multidimensional 26, 44, 45
92, 96

N

neuropeptides 56
new science 24, 81, 92
no-mind 20, 95
non-dual 17, 39, 44, 81,
105, 106

O

opiates 56
oxygenation 69

P

peak 44, 82
perception 17, 79, 95
physical blocks 77
plant proteins 65
polarization 24, 93
power of sound 46, 96
precision 38, 79
purification 81, 110
purificatory process 91

R

Reality 17
Recognitions Program 86
relaxation 34
rescripting 10, 51, 53, 60, 80
right-brain dominance 21

S

sacred dance 73
science of sound 46
sleep 29, 40, 60, 84
sound 102
soundtracks 53, 61, 100

Source Consciousness Awareness
18, 37, 96
spontaneous body movement 76
stereo systems 62

stress-inducing 40
stress-reliever 40
subconscious 31, 37, 38, 42, 76, 79, 93
subliminal affirmation 31
subliminals 51, 53
subtle bodies 26
support systems 64
supracausal 93
Synchronicity Experience 9, 44
Synchronicity Paradigm 10, 42, 44, 53, 86
Synchronicity School of High-Tech Meditation 112
Synchronicity Technology 86
synchronization 24, 29, 31, 71, 103
synchrony 44, 48, 52, 78
synchrony factor 86
synthetic perception 48
synthetically-induced highs 29

T

tape care 63
theta 36, 61
thought 17
transcendental access 28
transcendental bliss 29
transcendental states 54
transduction 93, 96
transformation 42, 53, 97, 104

U

unconscious 31, 37, 38, 76, 79

W

whole brain 10, 26, 27, 40
whole brain function 57
whole brain synchrony 28, 37, 56
witness consciousness 28